BIG BRAIN

BIG BRAIN

THE ORIGINS AND FUTURE OF HUMAN INTELLIGENCE

GARY LYNCH

AND

RICHARD GRANGER

Art by Cheryl Cotman

palgrave
macmillan

BIG BRAIN

Copyright © Gary Lynch and Richard Granger, 2008.

First published in 2008 by
PALGRAVE MACMILLAN™
175 Fifth Avenue, New York, N.Y. 10010 and
Houndmills, Basingstoke, Hampshire, England RG21 6XS
Companies and representatives throughout the world.

PALGRAVE MACMILLAN is the global academic imprint of the Palgrave Macmillan division of St. Martin's Press, LLC and of Palgrave Macmillan Ltd. Macmillan® is a registered trademark in the United States, United Kingdom and other countries. Palgrave is a registered trademark in the European Union and other countries.

ISBN-13: 978–1–4039–7978–0
ISBN-10: 1–4039–7978–2

R. Granger: (Figures 3.1, 8.1, 8.2, 8.3, 8.4, 8.5, 8.6, 8.7, 8.8, 9.1, 10.1, 10.2)
C. Cotman: (Figures 2.1, 2.2, 2.3, 4.1, 4.2, 5.1, 5.2, 5.3, 5.4, 6.1, 6.2, 6.3, 6.4, 7.1, 7.2, 7.3, 7.4, 12.2)
G. Lynch: (Figures 11.1, 11.2, 11.3, 11.4, 12.1)

Library of Congress Cataloging-in-Publication Data

Lynch, Gary.
    Big brain : the origins and future of human intelligence / Gary Lynch and Richard Granger.
        p. cm.
    Includes bibliographical references and index.
    ISBN 1–4039–7978–2—ISBN 1–4039–7979–0
        1. Brain—Evolution. 2. Intellect. 3. Neurosciences. I. Granger, Richard. II. Title.

QP376.L96 2008
612.8′2—dc22   2007036159

A catalogue record for this book is available from the British Library.

Design by Newgen Imaging Systems (P) Ltd., Chennai, India.

First edition: March 2008

10 9 8 7 6 5 4 3 2 1

Printed in the United States of America.

CONTENTS

CHAPTER 1

BIG BRAINS, BIGGER BRAINS

INTRODUCTION

Somewhere in Africa, sometime between five and six million years ago, began a process that led to an unprecedented outcome: the domination of the planet by a single species.

A typical mammal—a lion, a horse—has a world population of thousands to hundreds of thousands; but humans are now numbered in the billions. Typical mammals have locales, niches, in which they live: polar bears in ice, wolves in forests, apes in jungles. But we humans have broken out of our habitats and have fashioned almost the entire world into extended homes for ourselves. Animals kill other animals, for food and competition; but we humans manage to wipe out entire species, and kill ourselves by the thousands at a stroke.

And other animals communicate and even learn from each other, conveying apparent "cultural" knowledge. But no animal other than the human has any way to pass complex information to their great-great-grandchildren, nor can any other species learn from long-dead ancestors. Humans can do so, via language.

These differences all originate in one place. Species differ from each other in terms of bones, digestive systems, sensory organs, or other biological machinery; and until recently, these differences determined the winners and losers in the endless competition among life forms. But the human difference is as clear as it is enigmatic: it is our minds, and the brains that create those minds, that let us dwarf the abilities of other animals.

How did we get these brains, and how do they confer these unmatched capabilities?

These questions draw from many fields of study. *Biology* examines organs, from kidneys to pancreas; but brains are organs that uniquely produce not just biological but mental phenomena. *Neuroscience* studies brains; but brains are encoded and built by genetics, evolution, and development. *Psychology* studies the mind; but our minds are composed of our brains, their environments, our ability to learn, and our cultural surroundings. The countless facts and data compiled from scientific studies can overwhelm our understanding; but *computational science* synthesizes disparate facts, building them into coherent testable hypotheses; identifying candidate operating principles that may underlie the machinery of our brains.

This book marshals these disparate realms of scientific knowledge to ask how, a few million years ago, our ancient forebears began to grow brains far beyond their normal size; how the functions of those brains changed; and how that process led to who we are now.

The answers not only address what our brains can do, but what they cannot do. Humans build vast systems of roads, and vehicles, and power plants, but we struggle with their planning and their unexpected outcomes. We make scientific discoveries about our world, mastering mechanics, electricity, medicine; but it's not easy, and decades may go by between advances. Human societies develop complex economic and political organizations, but we barely understand them and often cannot control them. An understanding of how we arrived as the dominant creature on earth includes understanding our limits, the constraints on our mental powers . . . and glimpses of how we may overcome those constraints.

The path that led from ancient humans to ourselves is sometimes viewed as a relatively straight line of progress, from primitive to modern. We will show that it has been a path full of false starts and dead ends; of apparently aimless wandering interrupted by surprising leaps. Along the way, we will illustrate some of the remarkable turning points that led here, and introduce some of the ancient hominids who arose, and passed away, before we humans arrived.

Perhaps the most remarkable occurrence in our evolutionary history was the rise and fall of one of our very recent relatives. We'll introduce them, and use their similarities and differences as touchstones in our examination of ourselves. From the first discovery of their fossil skeletons and skulls, to reconstruction of their extraordinary brains, and inferences about their minds and their culture, their exceptional story will inform our own. For a time, they shared the earth with us; they walked the plains of southern Africa barely 10,000 to 30,000 years ago. They had most of our traits; they looked a lot like we do; and they were about our size . . . but their brains were far larger than our own.

BIGGEST BRAIN

Dr. Frederick W. FitzSimons had just been appointed the director of the Port Elizabeth Museum in 1906, and he took his new duties seriously. The little museum, which was tucked upstairs from the wool and produce markets in this small port town at the tip of South Africa, was in severe disrepair. "No real attempt at systematic classification, arrangement, or adequate labeling had hitherto been attempted," FitzSimons reported. "No efficient means had been taken to protect the specimens from the ravages of destructive insect pests."

FitzSimons closed the museum and had it thoroughly cleansed and refurbished. "I am pleased to state," he reported in 1907, that "I have completed the re-identification, classification, labeling, numbering and cataloguing," and that "further ridicule and criticism in regard to the state the Museum is in at present will be silenced." Upon its reopening, FitzSimons instituted a broad outreach program called "Popular Nights," offering public exhibitions, and featuring live snake shows. Local residents flocked to the performances, and the museum's reputation grew.

So Port Elizabeth Museum was naturally the place that came to mind in the autumn of 1913, when two farmers from the small inland town of Boskop dug up pieces of an odd-looking fossil skull on their land.

FitzSimons was as careful with these new bones as he had been with his museum. He quickly recognized that the specimens indeed formed a skull that was human, or strongly human-like; but he also realized what was most strange: the skull was simply too large. Neanderthal fossils had been found, with slightly bigger braincases, but this new one was huge. FitzSimons immediately saw that this was a stunning find: physical evidence of a human with a far bigger brain than our own. He performed a set of convincing measurements, and fired off a letter to the flagship science journal of the British Empire, *Nature*, describing the skull, noting its unique volume, and speculating about the heightened intelligence that would have come with its increased brain size. The skull's name, and the name of the heretofore unknown peoples that it represented, derived simply from the region in which it was found: these people were the Boskops.

The find was just as shocking to others as it had been to FitzSimons, and it didn't take long for the top anatomists and anthropologists of the world to get involved. Their subsequent examinations confirmed, and even extended, FitzSimons' initial estimate of the Boskop brain. Most estimates put the cranial capacity at 25 percent to 35 percent bigger than ours. Further digs were carried out in ensuing years, and more skulls, of equally superhuman size, were discovered. Neanderthal skulls, which had been discovered decades earlier, had large brain capacities but were shaped differently, with prominent bony ape-like brow ridges and less forehead than our own. But these new skulls had huge size along with fully human features. A human-like fossil in the Skhul caves in Qafzeh, Israel, had a brain capacity of roughly 1650 cubic centimeters,

20 percent larger than ours. Fossil skulls found at Wadjak in Indonesia, and at Fish Hoek in South Africa, each have 1600 cc capacities. Dozens of skulls from Europe, Asia, and Africa exhibit similar huge size, including familiar skulls that were found in the caves of Cro-Magnon, in southwestern France. (See table in Appendix). Boskops are the largest of them all, with estimated brain sizes of 1800 to 1900 ccs—more than 30 percent larger than ours. These brain cases have rising foreheads like our own, and have been found accompanied by slim, clearly human-like skeletons. The Boskops were around our size, between five and six feet tall. They walked upright. They had light, slender bones, and small, trim bodies—topped by very big brains.

Multiple scholarly articles were written about the Boskops and their brethren, and it became widely appreciated that a stunning discovery really had been made: previous humans had been bigger-brained, and likely smarter, than modern-day humans. Sir Arthur Keith, the most prominent anatomist in the British Empire, and president of the Royal Anthropological Institute, declared that Boskop "outrivals in brain volume any people of Europe, ancient or modern."

These discoveries caused a sensation in the early twentieth century. They were the subjects of conferences, the lead stories in newspapers, and were widely discussed in the scientific community. They raised a raft of questions: What does it mean to have a bigger brain? Are big brains definitely better? If so, how did their possessors die out while we *Homo sapiens* survived? Did they have brains that differed from ours, or did Boskops have the same abilities as we do? In particular, could they talk? Were they actually smarter? And . . . if they were such a big deal, why have most of us never heard of them?

ARE BIGGER BRAINS BETTER?

A human brain averages roughly 1350 cubic centimeters in volume, with normal brains easily ranging from 1100 to 1500 cc. From human to human, bigger isn't necessarily better: some very intelligent and accomplished people have small brains, and vice versa. At two extremes, satirist Jonathan Swift had an apparently giant brain of roughly 1900 cc, while equally noted writer Anatole France reportedly had a brain that barely topped 1000 cc. Geniuses are no exception. Einstein's brain reportedly measured an average and undistinguished 1230 cc.

For different members of the same species, a bigger brain may well be unimportant. But between different species, brain size can mean a lot.

Brains, like any other body part, are partly scaled to the overall body size of the animal. Bigger animals tend to have bigger brains, just as they have bigger eyes, feet, and bones. But some animals have features that don't seem to fit their overall body size: the neck of a giraffe, the teeth of a tiger, the trunk of an elephant. So if we measure the ratio of a body part to the overall body, most will maintain the normal size relations, while some will stand out from that scale.

On that scale, humans have normally-sized eyes, bones, and feet. But compared to other animals of our size, we have excessively huge brains. Our brains are smaller than an elephant's, but human brains are disproportionate: *for our body size* they are much larger than those of any other creature. Our nearest relatives are chimpanzees; if you take a chimp and a human of roughly equal body size, the person's brain, at roughly 1,350 ccs, will outweigh the chimp's brain by more than three times. For the same body mass, we have the equivalent of more than three of their brains.

This is unprecedented; if you chart the relation between brain size and body size, as we will in chapter 11, most animals will stay very close to the predicted ratios; humans will be wildly distant from them. One could argue that our brains are our defining feature, setting us apart from all other creatures in the world.

Indeed, it's our great brains, and our resulting intelligence, that changed everything in the world. Our vast population, our colonization of every corner of the earth, our remolding of physical features of the planet; all are new phenomena in a mere ten thousand years, after billions of years of life before humans.

What is it in our brains that led to our dominance, and what is it in Boskops' bigger brains that didn't?

BRAIN AND LANGUAGE

We can list the feats of intellect that differentiate us from all other animals: we make a dizzying array of tools, from saws and wrenches to wheels and engines; we heat cold places and cool down hot ones; we cook food; we travel huge distances around the globe; we build houses, roads, and bridges. These reflect many different abilities, but all are related by a hidden variable: our language ability. Ask yourself who it was, for instance, who discovered fire, or invented the saw, or the boat, or roads, or shoes. The reason it is very hard to answer these questions is that they were invented over time, by multiple individuals, who took what came before and improved upon it. The key here is that these unknown humans *built on what came before.*

Other animals interact with each other, and even learn from each other. There is evidence of other primates passing along "cultural" information and skills. But no animal other than us can pass on arbitrary information at will and across generations.

Dogs, whales, chimps, apes, don't have this advantage. Each is born to roughly the same world as their ancestors, and to make an invention they would have to do so themselves, within their lifetime. We have the unique ability to tell others something: something in addition to, beyond elemental, necessary skills. We can be told by our parents what a house is, what clothes are, what pencil and paper are; and in time we can tell our children, who can tell their children. Our individual brains take their jumping-off point from a mass of accumulated information that gets passed to us through language. Some chimps may have two parents, and a few teachers; language can give us the equivalent of thousands of teachers.

As an individual, a person may see a primitive boat and think of improvements to it—but as a group, we can pass that boat design on to many, and ensure that no individual will ever again have to re-invent it before improving it.

The process can fail. Indeed, such gaps have occurred in information transmission even within our short history of human culture. In the Renaissance, people saw the great domes of the Pantheon and other ancient buildings, which had been built a full thousand years before, and realized that they no longer possessed the ability to build such structures. The knowledge had been lost during the Middle Ages, after the fall of the Roman Empire and the concomitant loss of masses of written information and instructions. The Renaissance artists and engineers had to rediscover what the Romans had already known generations ago.

Language preserves knowledge outside of brains, and passes it from one brain to another. Whatever communicative abilities other animals have, our human languages have powerful characteristics that other animals don't possess. If we inherited our brains from our primate precursors, did they have some form of language? If they did, what did their language abilities look like?

And did the Boskops' bigger brains give them even greater powers of language? Or were their brains somehow deficient, bypassing the route to language? If they had it, why did they fail where we succeeded? If they didn't have it, why not, and what is it about our brains that gave us this ability?

WERE BOSKOPS SMARTER?

Brains are amazingly similar across all primates, from chimps to humans. Even the brains of dogs, and mice, and elephants, are all far, far more similar than they are different. We're all mammals, and the basic design of our brains was firmly laid down in the earliest mammalian ancestors, when they diverged from the reptiles more than 100 million years ago. The design has barely deviated since then, from the parts in a brain, and the patterns of their connections to each other, all the way down to the individual neurons that comprise them, and the detailed biochemistry of their operation.

But if the brains of a mouse, a monkey, a mammoth, and a human all contain the same brain designs, what are the differences?

Chimps are smarter than most animals, aren't they? Elephants have great memories, don't they? Dogs' sense of smell can sniff out minute clues better than other animals, can't they? We will show that these differences are actually extremely minor variations of the same underlying abilities. Chimps are smart because their brains are relatively large, not because those brains are different. Most mammals have the same great memories as elephants, whether or not we carefully test it. Dogs' keen sense of smell is shared by most mammals (though not us primates); we use dogs for tracking because we can train them.

And our own brains have most of these same designs and abilities. The primary difference, overwhelming all others, is size; compared to those of all other animals, our brains are many times too large for our bodies. With the great expansion of our brains came vast new territory to store immense tracts of memory, whose sheer extent changed the way we behave. Can it really be that changing the size alone can change its nature; that pure quantity can improve quality?

We'll show the small differences and vast similarities between ourselves and our primate relatives, and we will raise the question of size thresholds that may have to be passed for certain abilities to show up. Bring water to 99 degrees celsius, and it's hot water; raise it just one degree more, and it has new qualities. We will show what human brain changes look like, and explain the principles that enable them to occur. In general, larger mammalian brains show new abilities, as a dog outperforms a mouse, a chimp over a sloth, a human over an ape.

Or a Boskop over a human?

The evidence suggests that Boskops' brains were indeed very much like ours, only much larger; it strongly suggests that they would have been smarter than us. Their exact species is unknown. They may have been among our direct ancestors, in which case we seem to have devolved to our current smaller brain size, or they may have been a related, contemporaneous subspecies, our cousins; either way, it is likely that their substantial extra brain size would confer substantial added intelligence. Just as we're smarter than apes, they were probably smarter than us.

WHY HAVEN'T WE ALL HEARD
OF BOSKOPS?

Many of our hominid ancestors are almost household words: *Australopithecus*, *Homo erectus*, and of course the universally recognized Neanderthal, but Boskop is never mentioned.

As mentioned, the huge skulls from Fish Hoek, and Qafzeh, and Boskop, were at one time widely discussed. When these skulls were first found, they became famous indeed. All the top scientists studied them, and speculated about them. And they were widely known in the broader world outside of science; it was recognized that the Boskops were remarkable specimens, with strong implications for our history and our humanity. How did we forget them?

At the time of Boskops' discovery, Darwin's theories had already been published for fifty years, and had been widely accepted as part of the scientific canon. Evolution, although resisted by those who may have been offended by the suggestion that their ancestors swung from trees, at least had a comforting punchline: though we humans evolved from apes, we had evolved into something sublime, with powers unlike any other animal.

The Boskop skulls represented an impudent affront; a direct challenge to the presumed "upward" trajectory and ultimate supremacy of present-day humans.

Some researchers anticipated the reaction that would ensue. The Boskop discoveries would be attacked as either wrong or irrelevant. Evidence be damned; surely there could not have been smarter precursors of humans. Or, even if these skulls were irrefutably real, then perhaps the reasoning itself was in error. Even though our big brains clearly out-thought those of the apes, perhaps still bigger brains would not outthink our own. These emotional objections were presaged by the Scottish anthropologist Robert Broom, who wrote in 1925 to the journal *Nature*: "Prejudice has played a considerable part in anthropology. Since the belief in evolution became accepted, all old human skulls are expected to be ape-like, and if not ape-like are regarded with suspicion. . . . The Boskop skull has been threatened with a similar fate. It has an enormous brain and is not at all ape-like. Therefore, according to some, it cannot be old, and in any case it cannot be very interesting."

Broom accurately predicted that in the coming decades the Boskops would fall into obscurity. Part of the reason is just as Broom said: Boskop's brain is huge, and his brain and facial features are not at all ape-like, so he must be an anomaly. No one talks about such creatures, for they do not fit our ideas about who our ancestors were: cavemen one and all, brutish, lumbering, inferior. The Boskops were quite the contrary.

What does it mean? How did Boskops' supposed huge intelligence play out? How would it look to us, and how would it have felt to them? A Boskop's brain is to ours as our brains are to those of *Homo erectus*, an ancient caveman. We think of them as primitives; savages; how might the Boskops have viewed us?

OUTLINE OF THE BOOK

How can we pose, much less investigate, these questions? The Boskops are gone, and there's nothing out there with a bigger brain-to-body ratio than ourselves. But we can ask what it is in our brains that gives us intelligence, and more specifically what we have that chimps don't have, giving us the ability to plan, and to use complex tools, and language.

By analyzing the parts and interactions among the circuits of the brain, we can synthesize the ways in which they function during thinking. We can identify these key brain parts, and how they arose via evolution from primate ancestors to chimps on one hand and ourselves on the other. Armed with that knowledge, we can propose hypotheses of what new material the bigger brains of the Boskops would have contained. And, just as we can point to particular enlarged human brain areas and identify capabilities that they confer on us, we will make specific conjectures of the further abilities that the Boskops would likely have had.

We will spend time looking at skulls, but skulls alone won't do it: fossil skulls survive, but the brains within them don't. Scientists look at the space inside the skull, measuring it to find the size of the brain that occupied it. They can even look at the slight bulges and indentations, indicating the different extent of different regions on the surface of the brain; comparing those to similar living brains, they hypothesize the relative sizes of these different brain regions or lobes. Generating sweeping hypotheses from these skulls is hard. It has often been the case, for instance, that the inferences arrived at from analysis of skulls turns out to be at odds with the inferences that arise from analysis of genetic material. Which is right? The scientists involved often engage in extended verbal battles, sometimes lasting decades. Rather than picking sides in these debates, it's worth remembering that there is a right answer, even when we don't know it. It's not about who wins the argument; what's important is what the facts are at the end of the day. These simply aren't yet resolved, and throughout the book we will take pains to point out the remaining controversies, and the facts driving each of the different positions.

We'll study genes as well, but genes alone won't do it either: we have information about a number of genetic differences between ourselves and chimps, but still precious little knowledge of how it is that different genes yield different brains. We provide background on what is known of how genes build body parts, including brains, and we attempt to show some of the strength of the current state of knowledge, as well as the constraints on its interpretation.

We'll also use computation. Not computers, like a Mac or PC, but the underlying computational approach of describing the brain's operation in formal steps. Computational analysis can draw strength from two enterprises: the scientific aim of understanding the brain, and the engineering goal of building simulacra that imitate those mechanisms once they are understood. Most of the book will be about comprehending real biological brains, Boskops and our own. But we'll often veer to explicitly computational explanations, for two reasons. First, in order to illustrate when particular points about brains have been understood sufficiently well to

imitate them, that is, to test the theories in practice. Second
these insights to look into the future and to intimate what sc
on the cusp of achieving: new therapies, that may help fix brains
when they malfunction; and the development of brain machines,
that are capable of doing what we do uniquely well: thinking.

Our job in this book is to use knowledge of skulls, genes, brains, and
minds, those of ourselves, those of other extant animals such as chimps,
and even those of artificial creations such as robots, to extrapolate the
likely contents of the brains of Boskops, and, from their brains, to sur-
mise their mental lives. Along the way, we will notice striking, yet often
unexplored, facts about our own brains and our own abilities.

- A mind is what a brain does, and brain circuits are just circuits.
  As we can analyze the circuits in a TV or an iPhone, once we
  sufficiently understand the circuits in a brain, we'll be able to
  explain how they do what they do, diagnose their limits and defi-
  ciencies, possibly fix them when they break. This is at heart a
  computational understanding—not about computers, but about
  the computational functions of brains, and different functions of
  different brains (chapter 2).
- All the information used to build your brain and body is contained
  in your genes, and evolution changes genes. Many otherwise-
  confusing aspects of evolution are clarified by recognizing the
  constraints imposed by genetic organization (chapter 3).
- What's in a brain? We illustrate the pieces of brains, their origins,
  and how they interact (chapter 4). As brains grow, some parts
  grow huge; we describe in detail the largest parts of the human
  brain, dominated by the neocortex, and how it operates and
  learns (chapter 5). We propose the radical hypothesis that most
  of the cortex, and so most of the human brain, is designed around
  the olfactory system, the sense of smell, of ancient vertebrates. We
  show how the organization of those early systems came to be
  adapted to the brains we have today (chapter 6). What does the
  resulting system do? In particular, how does it go beyond simple
  perception and movement, to the internal processes of thinking?
  (chapter 7). As brains grow, they go from initial primitive

thought to human-level planning, reasoning, abstraction, and language. We describe the specifics of how high-level thought can arise from simple biological machinery (chapter 8).

- What makes one brain different from another? We show how the connectivity between brain areas determines the processing paths or "assembly lines" in the brain, and how subtle differences in the wiring of these brain paths can capture some of the primary abilities, talents, and shortcomings of individuals, and help explain the diversity among individuals and groups (chapter 9).

- What makes populations differ from each other? We ask the surprisingly difficult question of what makes a species and subspecies, what is meant by the notion of "race," and what the evidence is, and is not, for group and individual differences (chapter 10).

- Who are our ancestors, and what was their evolutionary path to us? We introduce the earliest hominids, our forefathers, and show the jumps that occurred in the sizes of their brains, and their abilities, over the past four million years (chapter 11).

- How did brains get to their enormous human size—and to the even-larger Boskop size? We describe the finding of hominid fossils, and their analysis, and mis-analysis. Some of the most celebrated fossils turned out to be frauds—how did these fool the experts? And more generally, how do differences of interpretation arise, and how can they be reconciled? (chapter 12).

- What are the detailed differences between the brains of humans and other primates and hominids, and how are they related? (chapter 13). Integrating these findings, we show how these hypotheses make speculative predictions about what the Boskops may have been like, and what we may become, as new biological and engineering technologies come into being (chapter 14).

We end the book with the questions that began it: What does it mean to be a big-brained human? Who were our bigger-brained ancestors of the recent past? Why did they die out? Why are we here; and where are we likely to go from here?

The Boskops coexisted with our *Homo sapiens* forebears. Just as we see the ancient *Homo erectus* as a savage primitive, Boskop may have viewed us somewhat the same way. It will be valuable for us to explore who they were; it will teach us about ourselves, and possibly teach us how we can be more than we are. And it will be worth investigating why they died out, while we remained and thrived. By learning their fate, perhaps we can avoid suffering it ourselves.

We shared the earth with the Boskops, and their bigger brains, for tens of thousands of years. This is a book about our huge brains, and the specter of the even larger brains that came before us. Time to learn a bit more about our betters, and about ourselves.

CHAPTER 2

THE MIND IN THE MACHINE

Fifty years ago, there was a conference for scientists who usually had nothing to say to each other. They came together to launch a scientific revolution.

The invitees were from wildly different fields: mathematics and psychology; biology and engineering, and some from the then-new fields of linguistics and computer science. They convened for a month on the idyllic campus of Dartmouth College, with the not-so-modest intention of starting a new area of research; a field in which their disparate disciplines would unite to solve some of the largest questions in science: what is thinking? what is intelligence? what is language? The mathematician John McCarthy coined a new term to describe the endeavor: "artificial intelligence." Initially obscure, it has become so widespread that it is now a proper topic for movies and blogs. Its name overemphasizes the "artificial"; it's really about understanding intelligence sufficiently well to imitate it. McCarthy put the goal succinctly: "to proceed on the basis of the conjecture that every aspect of learning or any other feature of intelligence can be so precisely described that a machine can be made to simulate it." To wit: if we can understand a brain, we can build one.

Building artificial brains can help us understand natural brains. Hypotheses about the brain are so complex that it is difficult to test them for their implications, or even for the internal self-consistency of the theories. If we can build even partial simulacra, we may gain insights into brain function. And such models may help us understand differences among different brains. Why are human brains so much more intelligent than the smaller brains of a chimp? What might larger brains, like Boskop's, be capable of?

Today, the idea of building brain-based computers isn't all that surprising. Computers already do all kinds of human-like things. They are, for example, getting pretty good at transcription: they'll listen to you (if you speak carefully) and copy down what you're saying. The Defense Department has computers that tap into phone conversations and recognize when suspicious words and phrases occur. Computer systems read parts of newspapers, and understand enough of what's going on in a story to recognize its potential impact on the stock market. And they are even starting to challenge professional poker players, acquiring a sense of that very human phenomenon called bluffing. But these types of operations are largely achieved with conventional machines operating at ever faster speeds with ever more clever programs. Can we go further than this, and build a machine that not only performs a few human-like functions, but actually acts like our brains?

Ongoing research at the interface between neuroscience and computation strongly suggests that it is possible to build silicon versions of brain structures—a momentous first step toward constructing an artificial brain. Scientists' understanding of both the biology and computational properties of brain circuits are steadily growing, and much of the book is about this progress.

LEARNING NETWORK CODES

We may want to build machines that share our mental abilities, but "mental abilities" are poorly defined; everyday terms for describing mental abilities don't actually *explain* those abilities. Since we all read, and think, and recognize, and understand, we intuitively think that we have explanations of *how* these tasks are carried out. But when trying to build a machine to read, or to recognize, or to understand, it becomes apparent that our definitions are shallow. An engineer building a bridge, or wiring up an iPod, knows exactly what these objects are meant to do, and so constructs them to carry out those specific functions. But for "recognizing" a face or "understanding" a news article, we have only observations of ourselves and others doing it, without internal specifications of what's going on in the machine, our brains, to accomplish the task.

Some might argue that a true appreciation of how humans deal with the world can be had from studying the mind rather than by focusing on the immense complexities of the brain. For decades, the mind was a quite separate topic of study from the brain, almost like a science of studying car behavior—accelerating, braking, weaving, parking—without ever lifting the hood to look at the engine. Perhaps in the end the mind can't be explained in terms of how the brain operates, and there are some who feel that current discoveries have already convincingly distinguished it from the brain. If so, the mind falls outside the scope of this book. We will take the standpoint that the mind is what the brain does; that minds can be understood by sufficiently understanding the brain. This is not to make a puerile reductionist argument, not to say that minds are "nothing more than" brains. Just as ecosystems are more than their individual components—oceans, forests, mountains, weather—and biological systems such as a kidney are more complex than any of their constituent chemistries, the mind arises from the interaction of multiple brain systems and their encounters with their environments. Studying brains in isolation won't give us the whole story of mental life, and studying them in context involves more than just neurobiology.

Rapid advances in neuroscience have provided a vast trove of often surprising results that can be applied to the problem of how the brain generates what we experience as thought. The next few chapters describe the background, and current state of the art, of these efforts to understand what's in a brain, and what it's doing when it listens, or recalls, or plans. These new findings are defining a new field of study, a field that finally maps the brain in sufficient detail to enable us to imitate it. It's a field that might best be called "brain engineering."

Early examples were forged back in the late 1980s. We and others were studying how brain systems behave as circuits; that is, describing wiring and functions in the brain from an engineering point of view. Early computer simulations of brain circuits turned out to perform surprisingly useful operations: some tasks that were hard for computers proved to be easy for these simulated brain circuits, such as recognizing difficult signals and sounds, from radar to EEG signals. It took years of further work before the artificial circuits began to gain the kind of power seen in real brain networks. To glimpse that power, it's instructive to tell a tale of deafness. Inside your ear is an organ called the cochlea, which takes sound waves and translates them into electrical pulses that are then transmitted inward to the brain. When scientists worked out the key mechanisms of the cochlea, they appreciated the intricate work it did to transform sound into complex signals. The cochlea was doing far more than just a microphone or a set of filters and amplifiers. In the spirit of John McCarthy, scientists set out to replicate the cochlea: to build an artificial cochlea that would work like the real one. In large measure, they succeeded in creating well-crafted and insightful silicon devices that carried out cochlear function. One of the obvious aims of the work was to construct prosthetics: implants that could help the deaf hear.

In many ways, this was a major departure from the standard approach to treating medical problems. Silicon devices are a relatively recent thing in medicine. Prior to this, if you were sick or injured, there were pills and surgery, not implants or reverse-engineered pieces of biology. But as scientists came to understand ever more biology, and began to imitate biological principles, they began to build prosthetics, like arms and legs, that were not rigid or inert, but could talk to the body, and listen to it, acting more like the limbs that they were replacing. The same was true for the ear: if you can build a device that does what cochleas do, then you should be able to wire it into the brain, where it could serve as a replacement part for a damaged cochlea. Researchers building silicon cochleas were thus on a revolutionary path to creating computational devices that might cure deafness. A remarkable story indeed. But one that gets even stranger.

At the same time that the electronic cochlea was being developed, other scientists were trying out a somewhat different strategy for treating deafness. In their view, the key aspect of a cochlea was not its elaborate and detailed processing, but its specific ability to selectively respond to particular frequency ranges. A hearing aid just turns up the volume on everything, and so amplifies much that is irrelevant to the listener. But what would happen if it were replaced with simple filtering devices that only enhance sounds that matter? After all, while silicon cochleas were remarkable engineering achievements, they were rare, and tricky, and expensive, whereas filters were well-understood, and small, and low-power, and cheap to produce. Sure enough, something remarkable happened. In patients who had lost their hearing, these filter banks were attached directly to the auditory nerve—the wire bundle that usually connects the cochlea to the brain. Initially, the patients simply heard noise, unintelligible squawking. But over the course of a few weeks, the patients got astonishingly better: they came to hear recognizable sounds, and in some cases even regained the ability to engage in conversation. A standard hearing aid was useless, but these implants were almost miraculous.

Why did the non-cochlear implants do so well? There are two important reasons. First, they actually captured one of the most important principles that underlies the cochlea: the selective and differential amplification of sound in different ranges, as opposed to hearing aids, which simply turned everything up. And the second point is even more telling. These implants worked because of what they were connected to. Remember that initially, they didn't work very well—it was only after some weeks that improvement occurred. What was happening in the interim? The implant itself didn't change; the brain circuits receiving the signals changed. They learned.

If you go to another country, or any region where people speak with a different accent, you initially have trouble understanding them. With time, you get better at it, and eventually the new speech patterns present no impediment at all. You learn the accent: your auditory brain circuits change subtly, translating the unfamiliar sounds to familiar ones. (You might even pick up a bit of the accent yourself—an important point we'll come back to later.) The same kind of thing was happening to the patients with ear implants: the electrical signals coming in were initially odd and difficult to interpret, but their brains learned to translate the sounds; to connect them up with the sounds that they knew well before losing their hearing. In other words, the implants were doing their part of the job, but the heavy lifting was being done downstream, not by the implant but by the patient's brain.

Implants are getting better and better, adopting more and more of the specialized processing of the cochlea—but still the primary work is being done by the receiving circuits, the brain that learns the codes.

Attempts to repair damaged vision are following a similar development path. Devices are being built to decipher simple shapes and movements, and these are being plugged into the nerves that go from the eye to the brain. The implants fall far short of the wondrous mechanisms in the retina. Instead, like auditory implants, they rely on the power of the brain. It is anticipated that the visual circuits in the brain will pick up the plug-in signals from the artificial eye implant, and learn to interpret them, possibly well enough to restore some measure of sight.

There are many details and caveats: devices of this kind are effective only in certain patients, especially those who already had hearing or vision, and lost it, rather than the congenitally deaf or blind; and their effectiveness varies substantially from patient to patient. But in each case, the key is that the peripheral circuits, substituting for ears and eyes, are only doing a fraction of the work. Their success is entirely dependent on the power of real brain circuits: the power to learn. Those internal circuits, those in the human neocortex, are the real prize.

What if we could imitate *those* circuits? Not peripheral circuits, like eyes and ears, but brain circuits that learn, that take inputs and figure out how to transform them into intelligible signals. It is these immensely complex learning machines, these brain circuit machines, that will be able to do what we do.

BRAIN CIRCUITS VS. COMPUTER CIRCUITS

Robots with artificial brains have been a staple of fiction for a surprisingly long time. They make their first appearance in a 1921 play by writer Karel Capek, called *R.U.R.*, or "Rossum's Universal Robots." The term was coined from the Czech word "robota" denoting forced work or manual labor. The play highlights and critiques the drudgery of robotic work, and presages the dangers that can arise: the robots in the play (actually biological entities, more like androids) eventually revolt against their human masters.

When we imagine robots, from HAL to the Terminator, we largely picture them acting like us. They scan the environment, store memories, make decisions, and act. Our brains enable us to do these things; theirs presumably would as well, whether constructed of silicon, or grown in vats. What designs do brains, natural or artificial, use that give them these powers?

While present-day computers are in some minor ways like brains, in most ways they're not, and the differences are profound. We can highlight five principles of brain circuits that set them apart from current computers: instruction, scaling, interactivity, integration, and continuity:

• Instruction: Learning vs. Programming
A brain can learn—by observation, or by being told. For instance, you can train your dog to obey simple commands, by repeating and rewarding. To get a computer to do anything, it must be painstakingly programmed; it can't be trained.

- Scaling: Adding Power vs. Diminishing Returns

Nature uses the same template to build the brains of hamsters and humans; but each brain naturally adds new abilities with size (e.g., from mice, to dogs, to apes, to people). We can build bigger computers, but their abilities don't change commensurately with their size. Today we have laptops with a hundred times the computational power of those from ten years ago, yet we're still largely running the same word processing and spreadsheet programs on them.

- Interactivity: Proactive vs. Reactive

Brains come with senses and effectors, ready to run a body. Computers come as a box, closed off from the world. They can use peripherals (cameras, robots), but these can be difficult to add and operate, and are not natural parts of the machine.

- Integration: Organizing vs. Depositing

You can see a bird, watch its flight, and hear its song, then combine these observations effortlessly into a concept, and you can immediately relate that concept to many others (other animals, other flying things, other singers). Computers simply deposit data into memory. No connections are built; no inferences are generated.

- Continuity: Memory vs. Blank Slate

Your previous experience is part of your behavior. What you did yesterday, and last week, changes you; not just learning from practice or mistakes, but also incorporating those experiences into your overall decision-making. Computers are the same every time you turn them on; brains aren't.

The potential benefits of building smart machines are clear. We'd be able to build robot workers, intelligent assistants, autonomous planetary explorers. We'd be able to build agents to perform either perfunctory labor or tasks that are terribly dangerous, or very expensive. Plenty of scientists have tried to build computers with these powers. Plenty have fallen short. The amount of time and money spent on trying to make smart machines is staggering; the military alone has funded programs amounting to billions of dollars, and industrial efforts have been undertaken on the same scale.

Given this history, it behooves us to pay attention to the one machine that can actually perform these tasks. If we want to build brains, we'll probably first have to really understand them.

The reciprocal is also true: to really understand brains, it helps immeasurably to try building them. There's nothing like an actual, working machine or computer program to test the internal consistency of your ideas. Over and over through history, it has proven possible to conceive of wondrous notions that didn't work when tried in the real world. Testing ideas by building them is a reliable way of finding the bugs in the idea—the little inconsistencies that are tremendously hard for us to figure out in the abstract.

The two enterprises go together: the scientific aim of understanding the brain, and the engineering goal of building one. Most of this book will be on the science side of comprehending real brains, Boskops and our own. But as we've said, we will often veer to the engineering side, in order to show when particular points about brains have been understood sufficiently well to imitate them, to test the theories in practice.

THE BRAIN OF JOHN VON NEUMANN

What is it about computers that differentiates them from us; that separates them in the five ways just listed? What gives them their powers (mathematical calculations, powerful searches, perfect memories) and their weaknesses (inability to recognize, to make associations between related facts, to learn from experience, to understand language)? Computers today rely to a surprising degree on the inventions of a single person. John Von Neumann was a true renaissance man, who made significant contributions to fields ranging from pure mathematics, to engineering, and to physics. Among other things, he participated in the Manhattan Project that constructed the first atomic bomb, working out key aspects of the physics in thermonuclear weapons. The work we will focus on was his design for early computers. He and others built on the ideas of Alan Turing to construct a "universal" computing machine, with a control unit and a memory. There are many interesting and intricate differences between Von Neumann, or "stored-program" computers, and other related systems such as the "Harvard architecture" which stores programs separately from data, but we will dwell instead on the far greater differences between all of these computer architectures, versus brain circuit architectures.

The separation of the control unit (or "CPU") and the memory unit cause computer function to be highly centralized; the CPU is the operational "bottleneck" through which every step must pass. Adding $2+2$, we might take the steps of storing a 2, storing another 2, performing the addition, and storing the result. Similarly, when you search the internet for a keyword or set of keywords, the computer has to search each possible site. Using multiple computers enables the task to be divided into parts, and all the results can then be combined into a single repository and sorted into the list you get back from Google. We might divide the work alphabetically, using twenty-six machines and giving each one a separate letter to search for. Or we might divide them by word length, with different computers searching for short, medium, and long words. Or by the geographic location of the computers on which the information resides, with separate searches for computers in each time zone. Some of these divisions make more sense than others, and it is not at all easy to divide one task into separate, parallel tasks in any useful way. So-called twin-core, and four-core, and eight-core computers add more CPUs acting in concert within a single machine. But except on very select tasks, they do not even approach being two times, or four times, or eight times faster than single-core machines. In general, if you add more processors, you get rapidly diminishing returns.

In contrast, the brain uses millions to billions of separate processors, and achieves processing speeds far beyond our current engineering capabilities.

A computer typically takes a terribly long time to run a visual recognition program, but brains, in their parallel fashion, will recognize a rose, a face, or a chair, in a fraction of a second. When an art critic recalls the Mona Lisa, she's activating millions of cells in the brain, and "assembling" the picture from those many parts. This architecture, involving thousands of independent engines all somehow acting in concert, is utterly opposite from the centralized Von Neumann processor of a computer. How a unitary image, or memory, emerges from so many separate operators is one of the great challenges of understanding brain operation. We construct a path to possible answers in later chapters of the book. The answers begin with attention to how neurons in the brain are connected to each other.

The circuit architectures within the brain come in two distinct flavors: point-to-point and random-access. These two kinds of connection patterns can be readily pictured by comparing analog and digital cameras. The analog version stores images as a pattern of tiny grains of light-sensitive silver, embedded in slightly gooey plastic. Each image stored on film is a direct replica of the visual image. You can physically look at the film and see an accurate point-to-point facsimile of the scene. Every location out in the scene appears in exactly its corresponding point-to-point location on the film: the house to the right of the tree, the boughs of the tree above its trunk. Point-to-point mapping is quite natural and intuitive.

But in a digital camera, the image is stored on a memory chip in the form of a very abstract encoding of ones and zeros; ons and offs. The codes are scattered through the chip. They are emphatically not laid out in any point-to-point fashion. To recapture the scene, the observer must apply a program, an algorithm, that *reconstructs* the image that has been secreted in the chip. The names of these codes are familiar—the images on the internet may be "jpegs" or "gifs" or "tiffs" or "pdfs". Each is its own, sometimes secret code. You can't view the code of one kind using the algorithms for another. No amount of staring at the chip will enable you to see the image; it is encoded, and must be decoded to be viewed. This is the general nature of "random access" mapping.

A comparable distinction is found in sound recording. A magnetic tape creates a direct point-to-point analog of the sequence of frequencies in a sound wave. Replaying the tape directly reproduces the sounds, and you can study it with an oscilloscope to see one-to-one correspondences mapped directly, point-to-point, between the replica and the sounds themselves. An iPod does something quite different. Frequencies and voltages are converted into digital encodings, again with familiar computer and internet names—MPEGs, MP3s—that have no resemblance to the sounds themselves. Again, they're encoded, and must be decoded. As with the camera, the digital sound recording device uses a method, an algorithm, to *rebuild* the original sounds, following computational instructions, algorithms, to decode the sounds from the internal stored ciphers. And again, the codes from one method can't be decoded with the algorithms from another. And, as with images, no amount of physical examination of the chip will extract the song. This is random access mapping of sounds, which, despite its current ubiquity, may seem indirect and counterintuitive in comparison with the more straightforward point-to-point method.

A brain has billions of parallel processors, in contrast to the small number of processors on a standard computer. And the brain's connectivity uses both forms of processing, and assigns them distinct roles within its architecture: some brain regions and systems use point-to-point design while others are hooked up in a random-access manner. In the former architecture, the connections maintain the arrangement, and the "neighbors" from one group of cells to another, thereby enabling the direct reproduction of an image, or a sound, as with camera film or magnetic tape. The latter architecture, random-access, connects cells in a complex, completely non-point-to-point manner. We can illustrate how these radically different designs are coordinated in a brain. Figure 2.1 depicts the body of a generic animal and the brain that controls it. Note that the various sensory inputs are nicely segregated on the body, and that this pattern is maintained in the brain. The nose at the front of the animal connects to the frontmost part of the brain. The eyes, a little further back, project to areas further back in the brain. The inner ear, processing both sounds and balance, find targets located another step back, behind the visual parts of the brain. Even the front-to-back axis of the whole body is mapped, front to back, onto the brain. The

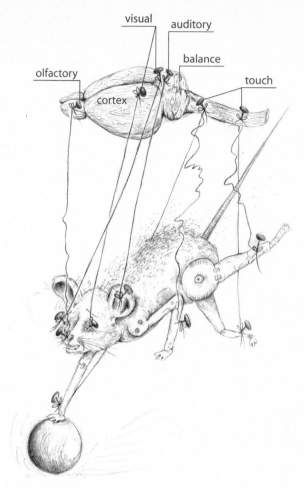

Figure 2.1   The organization of the brain parallels the organization of the body, so the nose projects to the front of the brain, the hind legs to the back, and eyes and ears to the middle. Each body region, or organ, has its own space in the brain.

animal doesn't send its sensory information to a central processor but instead sets up different regions for different modalities.

Going inside those carefully separated sensory regions, we find point-to-point connectivity patterns (see figure 2.2). And not once, but multiple times in serially connected relays. The retina projects point-to-point to a first stage, which then connects to a second stage in the same way, and so on all the way up the cerebral cortex, that vast final station sitting atop the brain. This is repeated for all of the sensory systems, with the great exception of olfaction (as we will describe in detail in the next two chapters). The cortex in this way winds up with physically separated analog maps of the visual field sampled by the retina, of the sound frequencies in a voice, of the skin surface of the body, and the muscles lying beneath it. But how to turn the complex patterns generated on any one of these maps by a particular stimulus, say a rose, into a unitary perception? Or, more mysterious still, take a pattern on the auditory map and combine it with one coming from the visual map? A rose after all can be correctly identified after hearing the word or seeing the image.

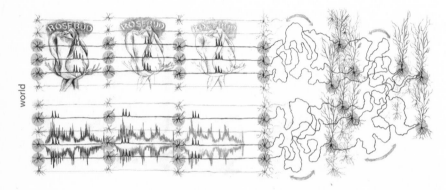

Figure 2.2   Messages from different senses (vision and hearing in the illustration) travel into their own brain regions where they are serially processed in a point-to-point fashion, so that replicas (albeit increasingly distorted ones) of real-world sights and sounds are momentarily created. From there, the patterns are sent into random-access networks where all organization is lost, and neurons randomly distributed throughout the network become activated (grey circles). Different senses ultimately merge their messages in a higher-order random-access network (right).

As shown in figure 2.2, the cortical maps send their information to subsequent areas, also in the cortex, in which the neurons are interconnected in the fashion we're calling random-access. In later chapters we'll describe the mechanism that lets these areas quickly and permanently alter their functional connectivity, enabling them to *encode a unitary representation of almost any complex pattern found in the point-to-point map regions*. And since these secondary, beyond-the-maps zones are all using the same random-access

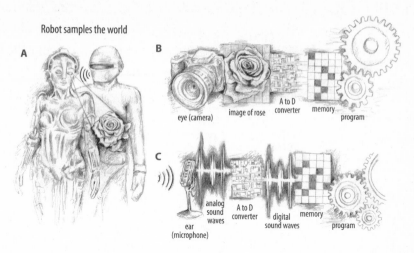

Figure 2.3 Present-day computers process sounds and images differently from brains. Computers directly code input into random-access memories without intermediate replicas of the rose or the voice. New kinds of computer circuits, derived from coordinated point-to-point and random-access brain maps, are the basis for novel robot brain systems ("brainbots").

language, they will have little problem connecting the representations assembled from the different types of maps. The image and the word, a certain touch and the memory of scene, are now combined.

Robots could be built with brains like this, combining these different internal styles of maps, point-to-point and random-access. The resulting "brainbots" might operate something like the fanciful illustration in figure 2.3. Visual and auditory cues, a rose and the sound of a voice, arrive at camera eyes and microphone ears where they are quickly converted into the same binary language. These signals are then stored on a memory disc in locations dictated by a program carefully prepared sometime in advance and controlled by a single CPU. The rose and the spoken words are now simply patterns, all but indistinguishable to an external observer, but internally denoting their separate meanings.

In these ways and in others, introduced in later chapters, we will see that the design of a brain diverges more and more from a Von Neumann machine. As we proceed, we will introduce the additional key aspects of this non-Von Neumann ("non-von") architecture and show how their differential expansion in big brains brings us closer to the origin of the human mind.

We will use insights from computation when we can, to illustrate the biological engines of our brains. We begin with the underlying biological systems that build brains: our genes. As we'll see, these are highly computational systems indeed.

CHAPTER 3

GENES BUILD BRAINS

We evolved from early apes; apes and all mammals evolved from reptiles; reptiles and amphibians evolved from fish. But how? No reptile woke up one morning and decided to become more mammal-like. Indeed, on the scale of a single individual animal, evolution is extraordinarily hard to understand. But the broader mechanisms of evolution, proposed by Darwin and Wallace and refined by many since then, can be understood by viewing each animal in two ways: first as the product of its genes, and second in terms of how those genes build bodies that interact with their environments.

It may appear that evolution strives resolutely forward, as though it were actively looking for new traits such as intelligence or language, or more empathy, or better short-term memory, to add to new species. In 1809, the French biologist Jean-Baptiste Lamarck (1744–1829), published a proposal that new traits acquired by individuals during their lifetimes, by practice or by learning, could then be passed on to their offspring. Lamarck's proposal—that your genes somehow pick up what you learn, and store it, and pass directly on to your children—is an attractive notion. If it were true, it would handily solve the most confusing aspect of evolution: how is it controlled; how does it seemingly become directed toward "more evolved" creatures? But Larmarckian hypotheses of "directed" evolution have never stood up to the evidence. The truth is stranger than Lamarck realized.

Evolution is predominantly an effect of absence, of laissez-faire: traits can be acquired via random, undirected accident, and if they happen to confer competitive advantage, or even if they merely *do not impair* competitive advantage, the traits may be passed on to offspring and retained in the species. In particular, random genetic changes can alter the brain, and if the result gives some individuals an improved ability to procreate, then that is all that's needed: any new traits will be more likely inherited by their progeny, who will be more likely to survive than those lacking the trait. No calculated response at all to the environment—just a set of blind trials that grope forward. Brains adapt by chance, and "most adapted" does not in any way mean "optimized"; it just means "able to scrape by" a little better than the next guy.

But these mechanisms for adaptation are like blunt instruments, blindly lumbering through evolutionary time. Such desultory processes seem utterly inadequate to explain the exquisite complexity of biological organisms. Surely fine-tuning is occurring; surely our bodies and brains are being somehow optimized. Somehow! The alternative seems ridiculous. How could random variation arrive at improvement? How can accidents turn reptiles into mammals, or apes into men?

We often fall into a fallacy of thinking—an almost irresistible fallacy—imagining that a feature or characteristic that we possess must have been carefully built that way, just for us. It's all too easy to believe that what is important to us—our hands, our faces, our ways of thinking—must also be important to evolution. It's crucial to remind ourselves that any organism alive today—a snail, a tree, a person—have all benefited from the same evolutionary mechanisms. Such creatures are not throwbacks; they're as evolved as we are. Evolution throws dice, tries out a possible configuration, and that configuration may thrive or die. This leads to dozens, thousands, millions of branches in our huge family tree, and each is a cousin, evolved in its own direction, adapting to its own niches.

The irresistible fallacy is to think that we *have* to be the way we are. But, in reality, we don't need to have five fingers: three or four would be fine; six would do well. We certainly don't need to have the same number of toes as we do fingers! It's just that the genes for one are yoked to those for the other, and there wasn't enough need to change them; we do fine with them as they are. We don't need to have hair on our chins but not on our foreheads. Our noses needn't be between our eyes and our mouths, and our ears needn't be on the sides. They'd work equally well, very possibly better, in slightly different configurations.

Throughout the book, we'll strive to point out where thinking sometimes falls into the irresistible fallacy, and we'll strive to catch ourselves when we too fall into it.

How, then, do our features evolve? How did we get five fingers and toes, and our eyes, ears, mouths, and brains?

The answer begins with genes. Evolution doesn't act on animals' bodies, but on their genes. Evolution doesn't turn reptiles into mammals—but it does turn a reptile's genes into proto-mammal genes, and those genes do the rest. What you're born with comes from your genes, and evolution changes genes.

Your entire body and brain are constructed predominantly of large molecules—and the instructions for producing these building materials, and assembling them into organs and organisms, are spelled out in your genes—your DNA.

DNA molecules contain within them the overall genetic blueprint for each type of organism, and each individual.

The various parts of DNA are named according to schemes determined in part by historical accident, as scientists were working to understand their nature. For instance: each unit of DNA is a "codon," which is a three-letter "word" that is spelled from an alphabet of only four "letters" or specific molecules. Each codon specifies the construction of a particular compound, an amino acid. These are the building blocks that make proteins, which in turn build the scaffolding of your body. Long sequences of codons form the "instructions" to grow proteins into structures, which deter-mine characteristics of an organism, including the shapes, sizes, colors, and copies of its various parts. Each such semi-independent sequence is what is referred to as a gene. Genes can be radically different lengths, from less than a thousand codons, to tens of thousands. Each separate strand or chromosome of DNA may contain from hundreds to thousands of these genes. The overall package of all of an organism's chromosomes is the full "genome."

It can be seen that some of these definitions contain ambiguities. In particular, genes can overlap with each other, can serve multiple functions in the same genome, can either produce proteins or can direct the production of proteins by other genes, and can occur in multiple different forms. A panoply of new, more specific terms has been introduced to refer to these different categories, but "gene" is still widely used, and we too will use it, in its relatively broad sense: a codon sequence whose products shape the formation of biological features.

HOW MUCH VARIATION CAN OCCUR?

A few key numbers help to visualize both the nature of genomic coding, and evolutionary variation in those codes.

Thinking of each gene as a document composed of words (codons), we can begin to count up possibilities. All codon words are spelled with an alphabet of just four letters (base pairs), three letters per word. There are sixty-four ways these base pairs can be combined into different codons, each of which specifies the construction of a specific protein component, or amino acid, of which there are only twenty, implying that several different codons specify the same amino acid; synonymous codon words for the same amino acid concept. In the language of DNA, there are but sixty-four words in the dictionary, and they can only say twenty things between them.

To compensate for this spare language of codons, genetic instructions use long, long sentences of them—up to tens of thousands of codons per gene. Following the instructions dictated in these genetic texts, proteins are assembled from up to thousands of constituent amino acids. From twenty amino acids, roughly 100,000 different types of proteins are created, are duplicated by the billions, and are assembled into a complete organism, all according to the rules in the genetic instruction manual.

But the first thing to note is that the genome seems too small for the job. A human genome is a personal library of classics containing about a billion codon words; the books in some people's houses contain that many words. Somehow, the complete construction kit for each individual is packed into each personal library.

For this reason, it was long thought that the number of genes (and codons) would increase with the apparent complexity of the organism. That is, a fruit fly would have far fewer genes than a human. Until it was possible for scientists to sketch out large gene maps, the numbers of genes in an organism were known only by estimates. Indeed, before the approximate layout of the human genome was worked out (by 2002), there were widespread bets among prominent scientists that it would contain more than 100,000 genes. These estimates were off—way off—by more than a factor of four: it turns out that we have roughly 25,000 genes.

The fact that a human being can be constructed from 25,000 genes is counterintuitive. A fruit fly has about 13,000 genes, perhaps half as many as we do. It's not easy to see how 13,000 genes makes a fruit fly, and just double that number somehow makes a human. Are all the differences between fruit flies and humans captured in a few thousand genes? Even worse (in terms of our pride as the dominant species), a mouse has about the same number of genes as humans. And so does a small flowering plant called a thale cress. And so do many other completely different organisms.

Recall that every gene is a different length, and the genes of mice contain slightly fewer codons on average than those of humans, so that an overall mouse genome is about 800 million codons whereas that of a human is a bit higher (about 900 million). The counts are still wildly at odds with our early intuitions: the genome of a lowly amoeba has been found to have more than 200 billion codons.

In sum, it is not the case that genome size grows in any way proportional to organism complexity.

And it gets more confounding the closer we look. The genomes of humans and chimps are reported to differ by just 2–3 percent, perhaps 400–500 genes out of 25,000, whereas the variability in the human genome itself has been revised drastically upward. It was long believed that all humans shared more than 99 percent of the genome sequence—i.e., that all humans differed from each other in at most about one half of 1 percent of their genomes. Experimental evidence now suggests that the genomic sequences of different human beings may differ from each other by as much as 12 percent.

These numbers seem not to make sense. We can change a human genome by as much as 12 percent, and still create a perfectly valid, different, human being. But we can change a human genome by as little as 3 percent and it becomes a chimp genome. It's not a paradox; it just matters *which* 3 percent gets changed. Instead of thinking of species differences by the amount of genetic difference, we must turn our attention to the specifics of *which* components are varied. When we tweak these particular genes, we get an opposable thumb. When we tweak those over there, we get bigger brains. If evolution tries out these genetic tweaks, it can arrive at the kinds of variations we actually see in animals. Inside the gene there are prepackaged instructions that make this possible.

BLUEPRINT SYSTEMS

Genetic instructions, when obeyed, construct complete working semi-autonomous systems—organs and organisms. The instructions are laid out more or less sequentially. They operate by being "read" by related mechanisms, transcription and translation, the central processes that read the DNA sequences, produce corresponding RNA, and then decode the RNA into amino acid sequences that constitute proteins. The resulting protein-based engines in turn perform all the complex tasks of an organism: digestion, locomotion, perception.

A cautious analogy can be made with computer software: roughly sequential instructions (computer codes), translated by related mechanisms (computer hardware and firmware), construct working semi-autonomous systems that can perform the complex tasks of a computing system, such as controlling a factory, operating the internet, running a robot. The analogy is a loose one: the details of the two systems differ enormously, and we will later encounter some quite-different (and equally imperfect) analogies of computers directly to brains, rather than computers to genes. For purposes of the present discussion, we guardedly note that both software and genes can be thought of as "blueprint systems," mechanisms that lay out the rules or blueprints by which complex machinery is built. These blueprints of course are not instructions to a contractor who interprets them—they are automatic blueprints, which run themselves without any intelligent, external intervention, and thus they must contain all the information typically included in not just a blueprint but also in the collective knowledge of contractor, builder, and carpenter. With this in mind, we will probe the similarities and differences of these two blueprint systems, genes and software. We will use the analogy for just one reason: to aid our understanding of genetic variation.

Consider the software that performs all flight control operations for NASA space shuttles, or the software that operates all Windows computers. The former contains approximately 2 million lines of computer code, and the latter more than 20 million. (We have no comment on whether one of these tasks actually is ten times harder than another, or whether some computer code is far more efficient than others.) For analogies with evolution, the question to ask is this: What happens when we make changes to this software?

Each individual line of computer code can carry out its own independent "instruction," telling the computer to perform a particular step. In these massive systems of millions of lines of code, if any individual line were to be randomly changed, the result would most likely be a program with a "bug"—that is, a program that doesn't work. In contrast, most changes to genetic material seem to generate new individuals—possessing slightly different traits, but all of them successfully living, breathing, digesting, moving, perceiving. Almost any random change to software produces a "bad" mutation, one that doesn't work, whereas genetic changes that produce bad mutations—individuals that are dysfunctional, or stillborn—are apparently far more rare.

Put differently, we can picture the task of productively modifying software, that is, creating new, useful individual variants of a NASA flight controller, or a Windows operating system. It is hard work—thousands of man-hours go into even small changes to these systems. By contrast, making genetic changes that result in viable variants of humankind can apparently be done via purely random variation! To understand evolution as "descent with variation," we must confront this counterintuitive puzzle: randomly varying computer code results mostly in errors, whereas randomly varying genetic code results in a rich repertoire of different, but equally viable, humans.

BUNDLING GENES

The strong mismatch between the way computer software responds to change, and the way genes respond to change, is crucial. It is the key to our understanding of how genes yield populations of species—which in turn illuminates the central question of how evolutionary mechanisms arrived at big brains. The key question is this: what is it about genetic codes that make them so much less brittle in response to change? Given how hard it is to create computer codes that can be flexibly modified without breaking, what underlies this ability in genes?

With ten-thousand-word texts, the twenty words of the genetic code can in principle be wrought into almost countless possible variants—far more potential variants than have ever actually occurred since the beginning of time. Yet genes do not and cannot actually generate all of these variants. Although the "alphabet" of the genome permits this vast array of possibilities, only a tiny fraction of them ever actually come to pass.

An analogy can readily be found in our own alphabet. The number of possible sequences of eight letters of the English alphabet is 26^8, or about 200 billion variants, yet we use only a tiny fraction of these. The eight-letter words that actually occur in English number fewer than 10,000—less than one ten-millionth of those that are possible.

Like letters, genetic elements are organized into preferred sequences, and longer genetic sentences are organized into "phrases" or modules, recurring in many different animals. For example, small evolutionarily conserved sequences called "motifs" often serve a related set of functions, or produce members of a class of proteins, as "prefab" components in many organisms.

Moreover, some of the phrases in genes go beyond the analogy to letters and words in a story. Some gene sequences might be thought of as "meta-phrases," or instructions directly to the reader (the transcription mechanisms) on how to interpret other phrases: whether to repeat them, ignore them, or modify them. Thus the same sequences are often re-used in different ways, being referenced by meta-phrases, as though we instructed you to re-read the previous sentence, and then to re-read it again skipping every third word.

The result is a complex genetic "toolkit" that includes, for instance, a few variants of body-pattern generators, such as the familiar pattern of a head, trunk, two arms, two legs; and related programs that have been well-tested over long evolutionary time, and that are re-used over and over in building an organism. The modules go a long way toward reducing variability in the gene sequences: for instance, variations can occur only in certain positions in a motif, but not elsewhere; and most variation occurs in the relationship among the modules, not within the modules themselves.

Indeed, software systems also use such strategies: computer scientists organize code into modular "subroutines" which can be separately tested and "debugged" so that they can then be inserted wholesale into much larger programs. And some code refers to or "calls" other code, meta-code instructing the machine how to operate on other parts of the program. These practices greatly improve software robustness, and many computer scientists suggest that more of this is better. Some go so far as to suggest that principles of gene sequence organization should be used to create software, which they hypothesize would be far less brittle. Anyone who has used a computer is aware of the fragility of software. It is often noted that if living organisms "crashed" like the Windows operating system does, they would of course not survive. As the secrets of genetic codes become increasingly understood, they may be taken up by software designers, enabling far more complex and robust designs than can currently be contemplated.

This modularity arises from, and contributes to, a key feature of genes: they are *compressed* encodings. They use a reduced "shorthand" to express well-worn motifs or modules that always play out the same way, not needing to have a rich, meticulously specified instruction set for the highly rehearsed "set pieces" that are used over and over across many different taxa of animals. In this shorthand, a brief message can denote a whole set of scripted steps. A gene can say simply "Bake a cake for two hours at 350 degrees," without having to say "break two eggs, beat, add milk, flour, sugar, stir, pour in cake pan," let alone having to say "walk to refrigerator, open refrigerator, remove egg carton, open egg carton, remove egg, break egg into bowl, discard eggshell, remove another egg, . . ." and so on. The longer the instructions, the more slight variations become possible (e.g., ". . . break egg into bowl, remove egg yolk"). Since the instructions in the blueprint are very short compared to the complexity of the organism they are building, those short instructions are stereotyped; always carried out the same way. New instructions can be substituted wholesale ("bake a pie" instead of "bake a cake"), but the internal instructions within the shorthand are highly limited in their modifiability. By and large, the whole "script" for cake-baking has to be run every time that instruction is seen.

Our experience with computer software gives us increased respect for the robustness of genes. To a computer programmer, it is almost incomprehensibly impressive that we can write the "program" for a human using just 20,000 parts, or using 20 million, or one billion. Indeed, researchers have been trying for many years to build software systems with the capabilities of humans in order to run robots and artificial intelligence systems, and have thus far found the task daunting. It is suggestive of a system that slowly worked out the bugs in low-level modules before proceeding on to use those modules in larger programs. It is even more remarkable to think that we could change a few lines of code here and there and get, instead of a failed computer program (or a stillborn organism), a fully functioning system with just slightly different behavior.

The organization of genes into toolkits or modules, then, makes genes far less brittle, and it does so by making them less variable to begin with. As in the example of eight-letter words in English, there are vast numbers of possibilities that never actually occur. There are a cosmic number of permutations that are possible when twenty codon words are organized into gene sentences of 1,000 words each, but if parts of the sentences are organized into immutable phrases, then most of the possible permutations will never occur. Figure 3.1 suggests the relative number of permutations, or individuals, that can come from changes (random mutations) that are made to a particular set of instructions of a given size. Part **a** depicts the relative number of variations that can be created when genetic instructions are not organized into phrases or modules, and it shows the relative percentage of errorful or damaged individuals that are likely to result from those random changes. Part **b** shows the two primary effects of organizing the instructions into modules. Modular instructions are restricted in the number of variations they can create, compared to their unorganized counterparts, but the "yield" from those mutations is much higher—that is, the varied individuals that are created are far more

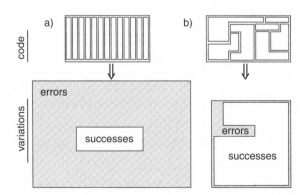

Figure 3.1 Genetic code organization produces different organisms. a) Non-modular codes (top left) can build many permutations of organism designs (bottom left), but most of these will be non-viable ("errors"). b) Organizing genetic instructions into modular packages (top right) reduces the number of possible variations (bottom right), but a greater percentage of the resulting organisms are viable ("successes").

likely to successfully survive. We hypothesize that this characteristic obtains for blueprint systems in general: that is, it holds for genetic codes and computer codes alike. Increasing modularity results in fewer variations, but a higher percentage of successful ones.

Figure 3.1. Effect of modularity on variation. Modifications to genetic codes (top) generate organisms (bottom) differently depending on the organization of the code. In nonmodular codes (a), many variants are possible but most of them produce nonviable mutations; modular codes (b) can produce relatively fewer variants overall, but a higher percentage of those variants are viable.

VARIATION IS RANDOM, BUT IT IS CONSTRAINED

Now let's look back at the process of random evolutionary variation. The earliest genetic codes may have been relatively unstructured, enabling vast possibilities of random variation, many of which produced animals that were likely unviable and quickly became extinct. If this were so, we would expect that early organisms existed in a profusion of wildly different forms, much more varied than extant animals in today's world. Such a hypothesis was forwarded by the late paleontologist Stephen Jay Gould, arguing that the now-famous Burgess Shale fossil site in Canada exhibits evidence of an extremely ancient (half a billion years old) trove of extraordinarily different and unexpected body types, a profusion of early variability exceeding that of current species. As certain patterns were stumbled upon that yielded functioning organisms, those patterns tended to be replicated in different species as nearly identical modules, even as the species themselves diverged.

It is hypothesized that there are certain types of genetic modifica-tions that are particularly adaptive: variants in which the codes are arranged modularly. That is, the internal organization into modules is itself likely to have been among the most useful adaptations, and is likely to have become increasingly dominant over evolutionary time. As modules slowly accreted into the genetic code, they limited the kinds of variants that could occur, reining in the initial broad variation; staying with patterns that sufficed and exploring variations only within the confines of those modular patterns. Species that evolved in this somewhat more conservative fashion tended to have competitive advantage over those that varied too wildly, and the trend toward modularization inexorably continued.

As a result, the number of possible genetic permutations is huge, but nowhere near as huge as it would be without this extensive structuring. There are presently a few dozen classes of animals, each with a relatively small set of body plans and chemistries. Some classes are especially regularized. All mammals, for instance, have a spinal cord with head at one end and tail at the other, four legs (sometimes differentiating between hind legs and forelimbs, which can be hands); all have two eyes and ears, one mouth; all have hair; all have highly similar circulatory, digestive, reproductive, and nervous systems. All variations occur within these (and many other) constraints. The constraints correspond to large components of our DNA that are shared, and remain unchanged with evolution of all reptilian and mammalian species. There are no mammals with a fifth leg growing from a forelimb elbow, or three heads, or tentacles, or sixteen eyes, or just one ear. Because genetic instructions are written in compressed modular shorthand, in practice only certain kinds of variations can ever occur—a few changes in the pre-packaged blueprints.

The resulting modular nature of variation leads us to a conjecture. Small random variation is occurring all the time—for every set of "unvarying" births within a species, there will, randomly, be a few variations created. Many of these may be either maladaptive, or insufficiently adaptive, or have linked side effects that render it maladaptive—one way or another, large numbers of variation attempts will likely fail. When some (relatively rare) adaptive variation does happen to occur, it will persist (by definition). In the fossil record, this pattern will show up clearly. The many small variants will have vanished without a trace (even if such small numbers did show up, they would rightly be rejected as aberrant individuals), whereas the rare successful (and persistent) variants will be seen in the record. Those rare variants will give the appearance of rapid or relatively abrupt changes, seemingly separated by periods of stability, during which variation appeared not to occur. This pattern is strongly suggestive of the very pattern that does occur in the fossil record, noted as far back as Darwin (1859; 1871), and labeled "punctuated equilibrium" by Eldredge and Gould (1972), though each of them had different accounts of it. Darwin attributed the gaps to losses in the fossil record, whereas Eldredge and Gould argued that geographic isolation and resulting "allopatric" speciation (arising from that isolation) were the primary factors. Others have taken still different positions—but this pattern of evolutionary variation is undisputed.

And so back to the brain. Within the strongly restricted blueprints for our body details, augmentations amount to little more than tiny incremental inventions. We can have slightly more versatile fingers. A slight modification of thumb placement can enable better manipulation. Variation in pigmentation genes can result in slight differences in the appearance of hair, eyes, skin. Adjustments of the hips let us walk on two legs. This conservatism in body plan is considered unsurprising, but brain changes are sometimes treated as open season for speculation, with theorists proposing the evolution of "new," quite different brain areas, specifically targeting new specialized behavioral faculties, almost magically arising to "respond" to environmental challenges.

We may think of bigger brains as good things; big brains make smarter animals, so surely evolution wants to increase brain size. But brains are expensive. Every cell in your body, including brain cells, require energy to operate. The reason we eat is to extract nutrients from other living things. We convert them into chemicals that fuel our cells like gasoline fuels a car. And it turns out that brain cells are the most expensive cells in your body, requiring approximately twice as much energy as other cells. Part of the cost is the expensive constant rebuilding of brain cells. Most cells in your body break down and are replaced over time, but the cells in your brain, with precious few exceptions, do not regenerate, and thus they have to engage in more laborious processes of in-place reconstruction. Just as it can be more expensive to extensively renovate an existing house than it would have been to build a new one from scratch, the upkeep on brain cells takes a great deal of energy.

The rest of the cost is expended by the unique job that brain cells do: sending messages throughout the brain via electrical impulses. This process, going on more or less constantly, is estimated to be responsible for about half the brain's energy expenditures—and amounts to almost 10 percent of all the energy expended by the entire body.

But biological systems have a strong tendency to shed anything they can, from parts to processes, during evolution. If a random mutation eliminates some expensive system, and the organism still thrives, then that organism may tend to get by with less food requirements than its competitors, and thus is likely to pass on its genes.

Given the tendency, then, to get rid of costly mechanisms, it is often seen as a wonder that human brain size has grown as much as it has. One might think that of all the parts of the body, the brain is the last that would yield an evolutionary increase. So the argument goes: if these highly expensive parts are being expanded, the results must be valuable indeed.

As a result, it is often hypothesized that each brain size increase during primate evolution must have been strongly selected for, i.e., there must have been some strong behavioral improvement that made the brain increase advantageous in the fight for survival. Hence the birth of fields from sociobiology to evolutionary psychology, which have generated strong hypotheses attempting to link behaviors to evolution. Such arguments include potential explanations for otherwise difficult-to-understand behaviors such as altruism, choice of mates, child rearing, and even language acquisition. Some argue that these are throwbacks to Lamarckian thinking; these traits are in us, so evolution must have put them there under pressure. But as we introduced earlier, we may be falling into the "irresistible fallacy" that all of our characteristics must have been carefully built just this way. Are these sociobiological arguments examples of the irresistible fallacy? Do these features really have to be just the way they happen to be?

We posit quite a different hypothesis: that brains increase for biological reasons—which may be largely accidental—and that behaviors follow this increase. As we have seen, it is relatively straightforward to posit how brain increase could arise from random genomic variation, but the question immediately arises how such a variation would be sustained in light of the added expense of a larger brain. This is where behavioral arguments often arise, and may indeed fit. It is not that a "need" or a "pressure" for a particular behavior, from sociology to linguistics, gave rise to a big brain. Rather, a big brain got randomly tossed up onto the table, and once there, it found utility. A randomly enlarged brain can find a previously-unexpected behavioral utility, and that utility may be sufficient to entail selection of the new brain size, despite its increased cost. A natural question is: what are the odds? If the enlarged brain is an accident, and the behaviors unexpected, how likely is it that a highly adaptive behavior set will arise from these accidents?

These questions set the topics for much of the rest of the book. We will examine questions of nature (innate abilities arising from genes) versus nurture (acquired abilities learned via interaction with the environment), in light of the sequence of events just described: (i) random brain size increase, (ii) unexpected behavioral utility, and (iii) maintenance of the big brain.

The basics have been established: evolution did not "figure out" that big brains would be useful, any more than it knew that slight displacement and rotation of a thumb would result in improved dexterity. Rather, modest and understandable gene variations stumbled onto these useful but relatively humble modifications. The genetic program for any mammalian brain remains almost entirely constant. It is likely that a few thousand kinds of changes, in just a few thousand modules of a few genes, give rise to all of the brains that occur in all mammals. Small genetic changes can trigger growth or reduction of body size, of limb size, and of brain size. In particular, as we will see, slightly longer or shorter gestation periods have a disproportionate effect on brain size, since most brain growth occurs very late in an infant's development.

Thus it is not just possible but highly likely that very small random genetic changes could have produced other hominid species, all of whom we are about to meet: Australopithecus, *Homo erectus*, Neanderthals, and Boskops, without optimization and without any particular fanfare—just as it subsequently gave rise to we "modern" humans, with a panoply of randomly toggled features, one of which was our big brain.

CHAPTER 4

BRAINS ARRIVE

In the earliest animals, brains began as simply a process of input and output: cells that linked a stimulus and a response. These initial "brains" are little more than collections of nerve cells, "neurons," processing inputs like light, sound, and touch, and producing outputs such as movements. Touch a snail and it will sense that touch (input) and it will retract into its shell, a reaction that involves sending a message (output) to its muscles. Not much information gets processed between the sensory input and motor output. In small brains, most of the work is dedicated to details of sensing inputs: touch, light, sound, smell, taste, and to producing outputs: various movements of muscles. The simpler the brain, the less material there is in between inputs and outputs. But as brain size increases, the proportion of it that is concerned directly with simple sensation (input) and movement (output) declines, and the more neurons there are in between. By the time we get to the size of a human brain, almost all the neuronal activity is entirely internal. Little is dedicated directly to the peripheral tasks of vision, or hearing, or other senses, or motor performance. Most of it is dedicated to thinking.

But all of the brain, periphery and "middle," is made of the same stuff: neurons, connected to each other.

Neurons are cells, like other cells in the body: skin cells, liver cells, etc. The difference is that neurons are specialized to send messages. They receive electrical inputs and send electrical outputs. They can be thought of a bit like simple little calculators, that add up their inputs, and send an electrical message as an output. The "messages" they send are just electrical pulses, and those brief transmissions contain no information other than their presence or absence: at any given moment they are either on or off.

A neuron gets a signal and it sends a signal, like a snail twitching to a touch. Neurons in your eyes get their signals from light itself—directly from photons striking them. Neurons in your ears get their signals directly from sound, from vibrations carried through the air. Neurons in your nose and on your tongue get their signals from chemical molecules that bind to them. And neurons in your skin are activated by the pressure of touch. In each case, a neuron receiving these inputs reacts by setting up an electrical signal. From that point on, all further signals are sent from neuron to neuron via electricity. And at the "output" side, they send an electrical signal to a muscle, which extends or contracts, moving part of your body.

FIRST BRAINS

Hagfish, and their lamprey relatives, jawless and slimy, look pretty hideous. Unfortunately for us, they are also our distant ancestors: the "stem" creatures that gave rise to all of the vertebrates—fish, amphibians, reptiles, birds, mammals. Those early ancestors had little in the way of brains—but the brains they did have, half a billion years ago, served as the basic design for all vertebrate brains ever since. Figure 4.1 is a cleaned-up sketch of the hagfish brain, as viewed from the side.

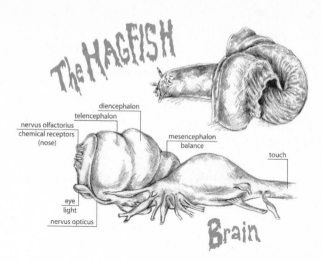

Figure 4.1   A hagfish (top right) is representative of the stem ancestors to all living vertebrates. Its brain (bottom) is largely an assemblage of sensory and motor organs. The forebrain (telencephalon), the area that constitutes most of the primate brain, is small in this primitive fish.

Sensory inputs. Our (necessarily brief) discussion of primitive brains will have to borrow results from scattered studies of several kinds of fish; it will use a number of inferences to paper over some pretty big holes in the literature. The hagfish central nervous system, going from the front (nose) to the back, is composed of an olfactory bulb, forebrain, diencephalon, midbrain, hindbrain, and spinal cord. This basic plan is found in all vertebrates.

Figure 4.2 adds some neurons to the picture. Neurons receive and send messages via electrical pulses, through wires that form their inputs and outputs. The input wires to a neuron are called dendrites and the outputs are axons. Axons from many neurons tend to become bundled together, traveling like an underground cable from one region of the brain to another. Nerves, like those in the frog's leg, connect our senses to the brain, and connect the brain to muscles. They are just thick bundles of axons coming from large groups of neurons.

Each input structure—olfaction, vision, touch—has neurons that send their nerves or axon bundles to specialized structures in the brain.

Nerve bundles from neurons in the nose travel to neurons in a first stage of the brain, the olfactory bulb, which in turn sends its axons to many subsequent stages throughout the forebrain. In these ancient stem animals, there's relatively little forebrain, and it is dominated by olfactory input, i.e., by information about smells. With each larger brain, the forebrain grows the most, until in humans it constitutes almost 90 percent of our brains.

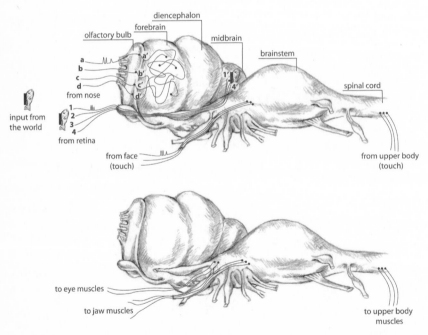

Figure 4.2 Nerve cells (neurons) arrayed in sensory and motor systems of a hagfish's body send connections via axons to neurons in its brain. (Top) The inputs are distributed in an orderly fashion, with those from the front of the animal (the nose) going to the front of the brain and those from the body surface landing in the spinal cord. Furthermore, neurons within an input (e.g., sight, smell) project in a point-to-point manner to their target regions, so the top of the receptor sheet inside the nose (a) goes to the top of the first stage of the brain (a' in the olfactory bulb). The olfactory system is unusual in that the second stage of processing (bulb to forebrain) is random. (Bottom) The outputs from the brain to muscles are also regionally organized.

Note that the projections from the nose to the bulb follow the point to point design, thus maintaining the organization found on the odor receptor sheet in the nose. But the connections from the bulb to the forebrain travel in a disorganized, almost random fashion. Here is an instance of the circuit designs we described in chapter 2. Distinct pathways arise from the eyes, and from the skin. From the eyes, neurons in the retina send their axons to neurons in an underlying area called the diencephalon; axons carrying touch information from the skin connect with neurons that are distributed along the spinal cord and associated areas in the hindbrain. The top of figure 4.2 illustrates how the locations of target regions line up with the corresponding position of sensory systems on the body: chemical smell sensors at the front of the animal activate frontal divisions of the brain (forebrain); the eyes, somewhat further back, send their axons to the next brain divisions in the sequence (diencephalon and part of the midbrain); the rest of the animal's body projects itself into the hindbrain and spinal cord. In general, each of our senses is a segregated operation with its own dedicated structures, with no central processor unifying them.

It's not surprising, then, that each brain area is located at a spot that corresponds to its inputs. More surprising is a continued correspondence within each of these areas. Within the vision area, there is a point-to-point map, akin to that in a film camera: inputs from the left part of the visual field activate neurons that in turn send their axons to a corresponding part of the diencephalon, whereas the right part of the visual field projects to a segregated region dedicated to the right-hand side, and so on for inputs that are in high versus low parts of the visual field. These point-to-point maps also occur for touch. There is a region of the hindbrain that is selectively activated when your hand is touched. It neighbors the region activated when your arm is touched, and so on. The result is much like a map drawn inside the brain, corresponding to locations on the skin, or in the image sensed by your eye, as seen in the figure. The resulting map in the brain is a direct point-to-point analog representation of the locations out in the world. These maps form naturally in the brain as an embryo grows to adulthood.

Motor outputs. The output side of the nervous system can be summarized quickly, in the bottom portion of figure 4.2. The hindbrain and spinal cord form a kind of motor column aligned with the map of body muscles. So neurons located at the very top edge of the spinal cord send axons to the face: to the nose, mouth, and eye muscles. Next in line, neurons near the front of the spinal cord project to the muscles of the upper body, just below the face. The pattern continues all the way down to the bottom of the spinal cord, which provides input to tail muscles. The spinal cord has masses of interconnected neurons in addition to those projecting out to the muscles, and these can, on their own, generate many of the sinusoidal body movements needed for swimming. The ancient wiring that generates those sinusoidal movements stays put throughout evolution. You can see it in reptiles; crocodiles winding their way across a mud flat—and it is still present, albeit greatly reduced, in mammals. Evolution is miserly, endlessly re-using and recycling its inventions like a parsimonious clockmaker.

In addition to the engines of locomotion in the hindbrain, there is a set of forebrain structures that is also critically involved in movement. This system, called the striatum or the striatal complex, acts like an organizer, globally coordinating the small movements of the hindbrain together into integrated actions. We will revisit the striatum in more detail in chapter 6.

These distinct motor regions are wildly different from each other. The types of neurons they engage, and the way those neurons are wired together, are strikingly different. Looking through a microscope at these brain structures, it's easy to immediately tell them apart. In fact, it's almost difficult to imagine that they are from the same brain.

Specialized movements each activate specialized brain machinery, calling on these areas in different sequence. To track a moving object, like a cat tracks a mouse, calls on direction sensors via the striatum, a stop-start pattern also in the striatum, and continuous body movement activated by the spinal cord. Much of the hagfish brain can be thought of as a collection of engines, each specialized according to different demands of the environment.

Sensory to motor connections. Inputs and outputs are relatively useless without each other. What good is seeing, if you can never act on what you see, and what good is movement that is undirected by information from the senses? As we mentioned, small vertebrate brains have barely anything in between the sensory inputs and motor outputs; larger brains contain disproportionately more and more of this middle material.

In a typical vertebrate brain, the visual areas of the midbrain, the neighboring auditory zones (when present), the cerebellum (when present), and the tactile areas all project into collections of neurons in the mid- and hindbrain that act as relays to the hindbrain-spinal cord motor column. There is considerable structure to all of this. The relays operated by the visual areas project to outputs aimed at the head, so as to orient the animal in the direction of a sensed cue. Analogously, touch areas connect to motor outputs that trigger muscle responses appropriate to the location of the stimulus on the body, so bumping into something with the front of the body is followed by one avoidance response whereas being grabbed by the tail evokes a very different reaction. The cerebellum's relays, when present, produce compensatory motor responses—if muscles on one side of the body are contracting, this system will make sure that other muscles are not trying to produce conflicting responses. In sum, there is a great deal of hard-wired, pre-packaged mechanism that comes built into even a primitive brain. It's a system with plenty of specializations, and pre-set point-to-point camera-film-like representations of the external world.

But there is one exception. One brain system stands in contrast to the rules that hold for touch, for vision, and for motor systems: the olfactory system. It has no hint of the point-to-point organization of the other systems. And indeed, intuitively, it's not clear how it could. In a visual image, we know what it means to say that the tree is to the right of the rock, and we can define the corresponding relation in the internal neuronal map. Analogously, it makes good sense to say that the touch to my head was above the touch to my arm, and that internal neuronal map is equally well defined. What corresponding map might we have for olfaction? Is a minty odor surrounded by a grass smell, or to the left or right of it? The locations out in the world don't stay put for olfaction as they do for vision. The top of a tree is above its trunk, but two smells might be in different positions on any given day. By moving around, we can tell where a smell is coming from, but smells themselves don't stand in any apparent relation to each other. If the odor system were like the visual system, we might expect the axon connections from the nose to the brain to build a point-to-point map that assigned different brain regions to different types of chemicals—floral, pungent, smoky, fruity, earthy—but no such map exists. Axons carrying olfactory signals distribute their messages almost at random across broad areas of the forebrain. The neurons that react to, say, a sugar molecule, are scattered across the surface of the olfactory system, with no evident relationship to one another. Unlike visual forebrain neurons, which have connections to corresponding hindbrain regions, the olfactory forebrain hardly connects at all to motor systems. It instead makes contact to other forebrain areas.

Big brains, including our human brains, retain this system for processing odors. But as we will see, the unique organization of this early sense of smell will come to form the basis of many more of our human brain circuits. Indeed, the unusual architecture of olfaction comes to form the first components of abstract thought.

BRAIN EXPANSION

Fish with jaws eventually evolved from the ancient stem ancestors of the hagfish, into a vast array of body types and lifestyles. Extraordinary modifications to the brain appeared in some of these lines, but through it all the same basic pattern was maintained. It took over 100 million years for the vertebrates to invade the land, and even then the move was half-hearted, in that the amphibians adopted a lifestyle that was only partially terrestrial. But shortly afterwards, reptiles, the first fully land-adapted vertebrates, emerged and flourished. Brains changed, particularly with the arrival of the reptiles—it is an easy matter to distinguish turtle vs. shark brains.

But the old pattern of bulb, forebrain, diencephalon, etc. is clearly still intact. But here is a truly surprising point: according to Harry Jerison in his classic book, *Evolution of the Brain and Intelligence*, the *relative size* of the brain does not change from fish through reptiles. Brains, like all organs, scale to body size according to well-established equations; Jerison's point is that the large groups of fish, amphibians, and reptiles kept to the same brain-to-body size equation laid down by the stem vertebrates. Individual species, such as sharks, may have gained larger brains, but big brains did not take hold as an evolutionary specialization. No species stood out for having an outsized brain with respect to its body. Animals from a goldfish to a Komodo dragon retained the same relative sizes of their brains to their bodies.

For hundreds of millions of years, then, the reptiles flourished on the earth, and the equation relating brain size to body size stayed constant. But that extraordinarily long period of brain size stability was about to end. The reptiles, having established themselves as the dominant large animals on land, split into a variety of subgroups, one of which, the therapsids, was destined eventually to become the forebears of the mammals. These proto-mammals seemed to have done well for themselves, apparently competing successfully with the reset of the reptiles. But they became challenged when a new reptilian offshoot arose. These new reptiles, the dinosaurs, had a whole battery of novel and fabulous adaptations, including bipedal locomotion: the first animals to walk on two legs. Dinosaurs were immensely successful species that quickly out-competed other reptiles and amphibians for most land niches, and even gained a few aquatic and aerial ones. It is not clear if this was accompanied by an increase in relative brain size. Using fossil skulls, Jerison estimated brain sizes for a number of dinosaurs and found them to fall comfortably within the ancient fish to reptile range, but others have provided evidence that certain dinosaur groups did in fact evolve abnormally large brains, including many familiar ones, such as the raptors that intelligently hunted in groups in Michael Crichton's *Jurassic Park*. We can get a hint by noting that birds are the only living descendants of the entire vast order of dinosaurs—and birds

have brains that are about three times larger than those of equiva-
lently sized reptiles. So quite possibly dinosaurs did indeed
increase their brain-to-body ratio, becoming the first giant-brained
vertebrates.

Meanwhile, as the dinosaurs came to rule the earth, what of the
once-successful proto-mammals, the therapsids?

Under what may have been intense pressure from their larger
reptilian relatives, the mammals-to-be began to change. These
nocturnal creatures began a path of almost reckless variation,
evolving a panoply of entirely new features, including hair, internal
temperature control (warm-bloodedness), breastfeeding of newborns, locomotion skills, such as climbing and swinging, that could
be used in trees, and two entirely new sensory specializations, one
dedicated to hearing and one to smell.

The specialization for hearing was a set of small bones in the jaw
that became modified into an inner ear. This structure amplifies
sounds and greatly increases auditory acuity; a powerful weapon in
the war with the dinosaurs.

The other sensory specialization, somewhat less appreciated,
occurs in the olfactory system. The proto-mammals evolved
turbinate bones: bony shelves inside the nasal cavity. These had the
effect of maintaining moisture and temperature of air as it slowly
passes across the internal surfaces of the nose. With this change, the
chemical odor sensors in the nose became better at detecting slight
distinctions among different smells. The number of these odor
detectors in the nose soon greatly expanded, and the mammals
became extensively olfactory creatures. That effect remains today: a
dog can have anywhere from hundreds of millions to billions of
odor detectors in its nose; that's more than the number of neurons
in the entire rest of their brain! In the transition from therapsids to
true mammals, then, the abilities to hear and to smell were both
markedly enhanced.

And, like the birds, the emerging mammals gained a brain about
three times larger than that found in their reptilian ancestors. When
we view this development in birds and mammals, we see an amazing case in which a particular exotic, radical adaptation appears in

the same geological time frame in two very different groups of animals. Relative brain size was steady for perhaps 300 million years and, for the majority of vertebrate species, has stayed that way up until the present. Then the birds and the mammals each created new classes of creatures, whose brains were far larger for their bodies than any before them.

The mystery deepens when it is realized that although both birds and mammals expanded their brains, their respective big brains were organized quite differently from one another. The division may have been based in part on how these two groups experience the world. Birds have fantastic visual systems. Eagles can spot a hare from a mile away, a feat well beyond the capabilities of any mammal. Dinosaurs such as Tyrannosaurus apparently possessed highly developed vision, and the earliest birds may have inherited and retained the already-excellent dinosaur visual system. When they took to the skies, their heightened visual abilities enabled them to navigate and to spot prey. Sight and smell, as discussed earlier, are processed very differently in the brain, and so it is reasonable to hypothesize that the expansion of different modalities—vision in birds, olfaction and audition in mammals—pushed brain evolution in two very different directions.

But we're doing it again: the irresistible fallacy. That tendency to think that the way things *are* is the way they were *pressured* to be. There are other reasons for why our senses evolved to their current levels. "The way things are" might be a side effect of some other adaptation. If so, then the side effects can result in new structural or functional features that were never themselves subject to selection pressures.

In the case of birds and mammals, it's possible that each came up with some other adaptation first, which then happened to enable the production of big brains. For instance, as noted both birds and mammals are warm-blooded. This comes with many advantages, including more energy to run the body, which can be seen in the relative activity levels of birds and mammals compared with their cold-blooded reptilian precursors. Big brains are actually selected against, since they add metabolic costs; that is, animals with bigger brains will tend to have a competitive disadvantage. But an animal with temperature regulation could absorb that cost much more easily than could a cold-blooded creature, and might be able to compete successfully despite the added metabolic cost. Thus the possibility of variants possessing big brains could occur with much higher probability in birds and in mammals than in reptiles. In other words, temperature regulation may have arisen first, and brains may have been more expandable as an unexpected byproduct. Unless these byproducts are overtly maladaptive, they stick around. We described these evolutionary selection processes in the previous chapter: as long as an accidental mutation doesn't hurt the reproductive survival potential of the animal, the mutation can remain as part of the phenotype, and becomes available to be further modified by subsequent evolution.

Competing with the dinosaurs, then, the early mammals evolved wildly, trying many new variations. Some of those, such as warm-bloodedness, may have generated more metabolic energy, and thus enabled brain expansion. When the dinosaurs disappeared, the mammals were unusually well-positioned to continue growing into the newly vacated niches, and wholly new functional possibilities began to open.

CHAPTER 5

THE BRAINS OF MAMMALS

While the avian brain was based on the reptile visual system, the new mammalian brain was based on reptilian olfaction, with its unique properties. We will outline the parts and characteristics of the reptilian olfactory system, sketch the way it works, and then show how that system became the template for the entire mammalian brain, including our human brains.

The primitive olfactory system in fish and reptiles has neurons arrayed within it in layers or sheets, like a set of blankets laid over other brain structures. In these ancient animals, this olfactory system is referred to as a "pallium," or cloak. When the first mammals developed, it is primarily this pallial structure that greatly expanded. The mammals grew and it extended until it covered much of the surface of their brains. The new structure is referred to as the cortex. The original reptilian olfactory cortex is transferred to mammals more or less intact, and all the rest of the mammal brain is the new cortex. We refer to the old, olfactory part of the cortex as "paleocortex," while the more recent parts of the cortex are called "neocortex." We will see that, as the brain grows, most of the growth occurs in the neocortex. We will often refer to all of it as simply cortex.

The new mammalian cortex thickens considerably, going from about a two-ply version in reptilian pallium and in olfactory paleocortex to a six-ply thickness, like multiple rugs stacked on each other, in neocortex.

In other, older brain structures, neurons are typically massed into clumps, in contrast to the cortex, arrayed in its layered carpets. Figure 5.1 emphasizes these differences, showing cross-sections through the forebrain of a mammal, exhibiting part of cortex, part of

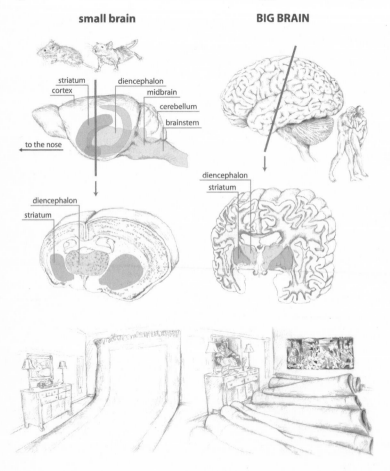

Figure 5.1   Big brains have different proportions than small ones. The brain of a small mammal (top left) has the same basic regions as the hagfish (figure 4.1) but the cortex has grown so large that it flops over the rest of the brain, covering some of the other subdivisions. A slice through the brain (middle left) reveals the clumped or "nuclear" structure of subcortical systems in contrast to the overlying layered "carpets" of the cortex. In a big-brained mammal (right), the cortex dwarfs the other brain divisions, and it takes on a crumpled appearance, folded into the skull like a carpet in a too-small room.

striatum, and part of the diencephalon. In sharp contrast to orderly layered cortex, neurons seem scattered uniformly throu_.. the striatum, and are grouped into clusters in the diencephalon.

As the cortex grew large, it grew uniformly: that is, throughout its vast extent it retains a repeated structure. It expands over the brain, and comes to take over the apparently different tasks of vision, hearing, and touch, yet throughout, it uses very much the same internal organization. We just saw that we can readily distinguish the striatum from the diencephalon, or the brainstem from the cerebellum, but it takes a skilled eye to tell one part of the cortex from another. Even looking at the cortex of different mammals, the designs are very difficult to tell apart. This kind of repetitive uniform design is alien to most of biology. For almost any other organ we find a collection of parts such as the compartments of a heart or kidney, each with specialized functions—functions that can often be deduced from their appearance and from their connections with other components. Not so for cortex.

NEURONS AND NETWORKS

Neurons are cells and large ones at that, and we can chemically treat them in such a way as to make them easily visible to the naked eye. The example in figure 5.2 illustrates the three primary parts of a neuron: its relatively small cell body, the massive dendritic tree rising up from the body, and a single, thin axon emerging from the base of the cell body. The axon can extend for great distances—all the way to the base of the spinal cord in some instances—giving off side branches as it goes.

The cortex is a vast forest of neurons. If a bird were the size of a small cell (say, a red blood cell as in figure 5.2), then a human cortex would be the size of the entire United States east of the Mississippi, entirely covered in an endless intertwined mass of dendritic trees above the planted neurons. As we've said, all a neuron does is communicate: it receives and sends simple electrical messages.

As described in chapter 4, all of that communication is accomplished through connections between neurons. Neurons send their output through wires, axons, to tens of thousands of other neurons. The actual contacts between a "transmitting" neuron's axons, and a

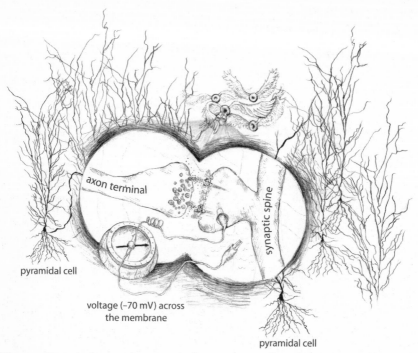

Figure 5.2 A forest of neurons interconnected by tiny synapses. A typical neuron (left) consists of extensively branching dendritic trees growing out of the tiny cell body, which sends its single axon out to make synaptic connections with the dendrites of other neurons. The cell body generates an electrical pulse that travels down the axon until it reaches the 'terminal', at which point chemical transmitter molecules are released. These cross the tiny synaptic gap towards a 'spine' on the dentritic branches of another neuron, inducing a new electric current. Unlike the drawing, the cortex is actually densely packed with neurons whose dentritic trees are endlessly intertwined. If small cells were the size of birds (top), they would look down on cortex and see the canopy of a vast, dense forest stretching far beyond the horizon in every direction.

"receiving" neuron's dendrites, are synapses: sticky junctions on the twigs of the limbs of target dendritic trees. Each synaptic tree can have tens of thousands of these contacts from other neurons.

The electrical pulse traveling down the axon causes the synapse to release a chemical neurotransmitter, that then crosses a very thin space to reach the dendritic spine; 2,000 of these tiny gaps would fit comfortably inside the thinnest of human hairs. The chemical neurotransmitter for almost all synapses in cortex is a small molecule called glutamate made up of about 30 atoms. A glutamate neurotransmitter molecule binds to the synapse at a "docking" site called a receptor. The binding of the transmitter causes a new electrical signal to be induced in the target neuron.

Of course, these messages from neuron to neuron, sent through synapses, only work if the synapses do their job. Yet synapses are notoriously unreliable. Any given input message will release the glutamate neurotransmitter only about half the time, and reliable thoughts and actions arise only from the co-occurrence of thousands of these iffy events. These neurons and synapses are components of a circuit, but they are not components that any self-respecting engineer would ever choose. If a chip designer at Intel used connections of this kind, he'd be summarily fired. Yet the brain uses these components, and as we've seen, the brain can perform tasks like recognition, that computers can't. As we discussed in chapter 2, one of the great mysteries we'll address is how the mediocre components in a brain can perform together to generate machines that perform so well.

The design of the olfactory system, your smell system, can be viewed in three parts: (1) cells in the nose, which connect to a structure called (2) the olfactory bulb, which in turn connects to (3) the olfactory cortex (figure 5.3). This system exhibits both of the circuit organizations we have been discussing. The axons from the nose to the bulb are organized as point-to-point connections, so that any pattern of activity activated by a particular odor in the nose, is faithfully replicated in the olfactory bulb. But the connections from the bulb to the cortex are the other kind: random-access circuits. So an olfactory pattern, carefully maintained from nose to bulb, is tossed away in the path from bulb to cortex.

This scrambling of the message is the key to its operation. Why would a circuit start with point-to-point information about which specific areas of the nose were active, and then throw the information away, transferring it in random-access manner to the cortex, where the signal can connect anywhere at all without organization?

As we've discussed, these random-access designs in the olfactory system can be understood by reference to the nature of odors and their combinations. You can buy a candle that contains coconut, pineapple, citrus, ginger and sandalwood; or a soap emitting peach, bay leaf, and rum perfume. Each of these combinations can produce a single unified olfactory experience: you can recognize the smell of that candle or of that soap without having to mentally step through all the separate ingredients. What random-access networks do is enable unified perceptions of disparate ingredients.

In figure 5.3, axons from three separate sites in the olfactory bulb (which reflect three corresponding sites in the nose), travel through the olfactory cortex in random-access mode—and every so often, they arrive at cortex neurons on which they all converge. These converged-upon targets are very strongly activated by this pattern in the bulb, since the cortical neurons are receiving three simultaneous electrical messages, whereas other neurons, even close neighbors, may receive only one or none. Thus these targets become "recruited" to respond to this particular odor combination. The beauty of random-access connectivity is that it enables individual target neurons to be selected in this fashion, and assigned to a novel odor composed of any arbitrary constituents. The recruited target neurons, in a sense, become the "name" for that new odor. In point-to-point connections, as in initial vision circuits, this can't happen. If there are different parts to a visual input pattern, e.g., the lower trunk and upper branches in the shape of a tree, there will be corresponding parts to the output pattern. But in random-access circuits, those same parts of an input, the visual images of a trunk and boughs, may select a single target output that denotes the simultaneous occurrence of all the parts of the tree pattern, rather than its separate parts in isolation; thus the responding neurons denote the whole tree, rather than just its parts. That is, a target neuron that responds to all parts of the tree input, converging from various parts of the trunk and branches, is acting as a recognizer, a detector, of the overall tree pattern. As described in chapter 2, the target neurons in a random-access circuit have the ability to detect these gestalt-like patterns—patterns of the whole, not just the parts. Each neuron in a point-to-point network can only respond to its assigned isolated parts of the input, but each neuron in a random-access circuit may, all by itself, respond to aspects of the entire input pattern. Different cells in a random-access circuit lie in wait for their particular combinatorial pattern—an input arrangement from any parts of the scene to which they happen to be very well connected.

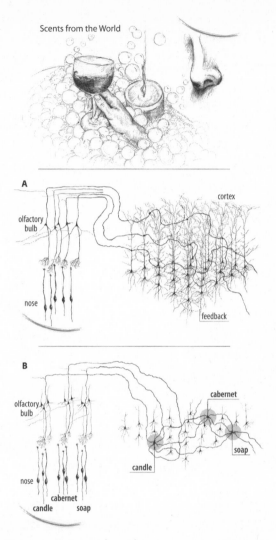

Figure 5.3 Point to point and random-access organization in the olfactory system. Arrays of neurons in the nose (A) send axons to the olfactory bulb, maintaining their spatial organization in point to point fashion. But the bulb then discards this organization, sending axons in diffuse, random-access patterns to the olfactory cortex (far right). The axons of these neurons have branches that double back into the olfactory cortex, further mixing up the already mixed-up input signals. Even though three different aroma components may activate three different locations in your nose, and in the olfactory cortex, further mixing up the already mixed-up input signals. Even though three different aroma components (B) may activate three different locations in your nose, and in the olfactory bulb, their random-access projections will intersect at various random points in cortex: cells at which the inputs converge can be used as storage sites for a unified percept (e.g., "candle" or "cabernet" or "soap").

The great advantage of random-access circuits is that there will be *some* scattered population of cells in random-access cortex that respond to *any* possible combination of odorants, including combinations that are entirely unexpected, and that may never have occurred before. Random-access circuits solve the problem of how to assemble a diverse and unpredictable collection of inputs into a unitary and unique output.

Suppose an animal's nose has receptors for 500 different odorants. Any small collection of these odorants might combine to make a real-world smell. The number of *possible* smells reaches far past the billions, and it is the random-access circuit design that can accommodate all of these combinatorial possibilities.

LEARNING

Perhaps the single most crucial feature of random-access circuits is this: they can be modified by experience. The more a particular set of connections are activated, the more they are strengthened, becoming increasingly reliable responders. The way they do this is worthy of a book in its own right—but suffice it to say that it is one of those instances in biology in which a broad swath of observations all fit together with amazingly tight coordination. When a mouse, for example, sniffs an odor, she sniffs rhythmically, about five times per second. It's not voluntary: she's biologically wired to sniff at this rate. The receptor cells in the nose, and then in the olfactory bulb (figure 5.3), and then the olfactory cortex, are all activated at the same frequency. When the activation reaches the cortex, something remarkable happens: the synaptic connections in the cortex contain a biological machine that can permanently amplify the signal, strengthening the synapses; and that machine is selectively activated by the precise rhythmic pattern of activation triggered by sniffing. In other words, when the animal is actively exploring her environment, the resulting brain activity causes synaptic connections to strengthen, enabling her brain cells to respond more strongly to this odor in the future. When she explores, she sniffs; and when she sniffs, she learns. All the components, from the submicroscopic world of proteins, voltages, and exotic chemistries, are

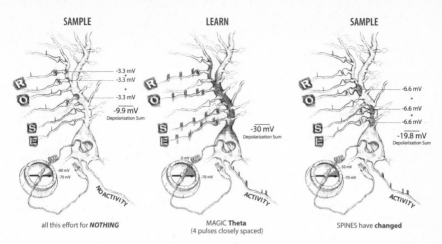

Figure 5.4 How memories are encoded. Three stages of learning are illustrated. (A) An input (R.O.S.E.) activates four axons; because release is probabilistic, and often fails altogether, transmitter (small black dots) is shown as coming out of only three of four the active axons. The released transmitter causes voltage changes (−3.3 mV in this example: mV is 1/1000th of a volt) at three spines (small extensions of the dendritic tree), resulting in a total 9.9 mV drop for the entire cell, as recorded by a voltmeter with one input inside the neuron and the other immediately outside. The total voltage change is not sufficient to cause the cell to send a voltage pulse down its axon ('no activity'). (B) Learning happens when the inputs are driven in a 'theta' characteristic pattern (note the little voltage pulses on the axons). The learning pattern overcomes probabilities and causes all synapses to repeatedly release their transmitter. Combined, these events produce a large voltage change (−30 mV) in the neuron; the active (transmitter released) synapses grow stronger when these events are packed into a very short time period (less than 1/10th of a second). (C) The original R.O.S.E. input signal is the same as it was before learning but now the released transmitter lands on modified spines. Because of this, the voltage changes are doubled, resulting in a summed value that is great enough to cause the neuron to react and send a voltage pulse down its axon. This cell now 'recognizes' the word 'rose'.

all tightly linked to this rhythmic activation pattern. The mechanism for this had its origin more than half a billion years ago, in the rhythmic tail movements of primitive fish that enable swimming. When you're learning a new telephone number, you're engaging some of the same biological processes used by a hagfish slithering through the water. Biology uses and reuses its inventions, retaining and adapting them to new uses.

The ability to strengthen synaptic connections is a simple form of learning. Strengthening a connection simply locks in a particular pattern in the brain, and that responding pattern becomes the brain's internal "code" for whatever is being sensed. Your experience of a chocolate chip cookie is not a photograph or a recording; not a camera image, but an internal construct, a cortical creation.

Once the cortex developed from the olfactory precursor circuits, it became independent; cortex no longer had to operate solely on olfaction, its original mode. In mammals it became the engine that analyzes olfactory odors, *and* visual images, *and* auditory sounds, *and* the sensation of touch, and a great deal more. And as we've said, how the nearly uniform structures of cortex can end up doing the very different jobs of the different senses is another of the major mysteries of the brain.

Images and sounds are converted into internal random-access codes much like olfactory codes. As these codes are created, they can be readily transmitted downstream to any other brain area, all of which now use the same internal coding scheme. And, using those connections, two different senses can be directly hooked together: the smell of the chocolate-chip cookie and its shape; its taste; the sound when it breaks. Using the shared cortical design, every sense acquires the same capability. The sound of a song can remind you of the band that plays it, the cover art on their CD, the concert where you saw them play, and whom you were with. All these senses participate in perceiving the event, creating memories of the event, and retrieving those memories.

Our mammalian brains, organized on the olfactory random-access design, give us these abilities. This represents a great divide in the animal world. Reptiles and birds have only their relatively small olfactory systems organized in random-access circuits. Mammals took that minor system and massively expanded and elaborated it. For reptiles and birds, it's an entirely different prospect to transfer information between their point-to-point designs for images and sounds. For mammals, it's natural.

These random-access circuits became a template for the explosive growth of cortex as mammal brains grew ever larger. Not only do these cortical circuits now operate not just olfaction but also vision, touch and hearing; they also generate the rest of the mental abilities in the mammalian brain. Our most sophisticated cognitive abilities are still based on that ancient design. Adding more of these same structures generates new animals with new mental capabilities.

CHAPTER 6

FROM OLFACTION TO COGNITION

The new cortical circuits that arose with the first mammals are the beginnings of our human intelligence. But cortical circuits emphatically do not operate in isolation; they are tightly connected to a set of four crucial brain structures just beneath the cortex. Each of these four subcortical systems, while working in concert with the cortex, carries out its own operations, conferring specific abilities to the overall operation of survival.

As we will see, these different subcortical components can be thought of in terms of a small set of questions that they answer for the cortex. When the cortex "recognizes" the presence of an odor, questions rapidly arise. Is the odor familiar or unfamiliar? Is it reminiscent of other odors? Is the odor dangerous, or attractive? Has it been associated with good or bad outcomes? Rewards or punishments? Does this odor tend to occur at a certain place, or under certain conditions, or together with certain objects? What other events were occurring when this odor was previously encountered? Are there actions that should be taken, from simple approach or avoidance to complex tracking or planning?

These questions are as ancient as the machinery that addresses them. Pathways emanate from the cortex to the four primary brain circuits that are responsible for tackling them.

Figure 6.1 is a drawing of a generalized mammalian brain showing these four major targets of the olfactory cortex: striatum, amygdala, hippocampus, and thalamus. Each is very different from the others—different circuit structure, different connections, different functions. All connect with cortex. As we will see, they work with cortex to control not just our behaviors and reactions, but our thoughts, our decisions, and our memories.

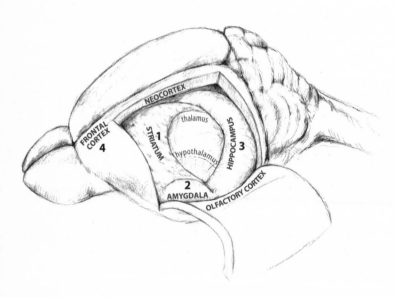

Figure 6.1 Organization of the four very different subcortical targets to which the olfactory cortex connects: 1) striatum, 2) amygdala, 3) hippocampus, and 4) frontal thalamo-cortical system. Each combines with other structures to carry out different specialized processes during perception, learning, remembering, and planning.

Striatum. This structure, briefly introduced in chapter 4, sends its outputs to brainstem areas connected to muscles and the spinal cord. It is clearly involved in getting the body to move. The striatum can be understood by picturing its outputs. In reptiles, and in most mammals, the striatum sends its messages to the ancient hindbrain and brainstem structures that control muscles. The brainstem systems each produce small component behaviors, like a twitch, which comprise the set of primitive movements that the animal can carry out. What the striatum does is play these separate brainstem components, activating neurons like keys on a piano, constructing whole musical tunes and harmonies from the individual notes.

The striatum has become wonderfully well-understood in recent years, giving over its secrets to a generation of determined scientists. The basics are these: the striatum contains circuits that activate, and others that suppress, the hindbrain muscle systems. Through these two networks, a message sent from cortex to striatum can initiate a "go" signal or a "stop." Thus an odor recognized in the cortex can set the animal in motion (e.g., in response to food) or cause it to freeze (sensing a predator).

The different functions of the hindbrain and the striatum can be readily seen in experiments. A brain scientist can touch an electrode to a hindbrain region, and this will evoke a circumscribed, jerky motion in some muscle group (depending on exactly where the electrode is). But touching the electrode to a part of the striatum will instead evoke a coherent, organized sequence of motions, played out by the controlling striatum.

One-time Yale professor Jose Manuel Rodriguez Delgado staged perhaps the most famous demonstration of the efficacy of this "electrical mind control." At a bull-breeding ranch in Spain, in the 1960s he implanted radio receivers into the brains of several bulls. He then stood unarmed in the middle of a bullfighting ring, and set the bulls loose, one at a time. By pressing buttons on his hand-held transmitter, he virtually controlled the animals' behavior. At one point, a bull was charging directly at Delgado; pushing a button, he caused the bull to skid to a stop and turn away. The entire episode, captured on film, was covered by the news media worldwide. A front-page story in the *New York Times* called the event "the most spectacular demonstration ever performed of the deliberate modification of animal behavior through external control of the brain."

(The striatum was also the subject of perhaps the greatest case of misidentification in brain science history. Anatomists, up through the middle of the twentieth century, had trouble finding it in birds and wound up convincing themselves that it was every-where, that most of the avian forebrain was in fact a hyperdevel-oped striatum. Given the links between striatum and movements in mammals, the presence of a colossal striatum led to the fasci-nating deduction that birds have enormous sets of locomotor programs, and therefore that their brain is a giant reflex machine. It was eventually recognized that birds actually have a reasonable-sized striatum on top of which sits a much larger region, just as in the case of mammals. In the wake of this revelation, the field of avian and mammalian comparative neuroanatomy is currently undergoing a huge upheaval; in 2004 and 2005, a series of papers were published proposing a complete re-naming of almost every major structure in the avian brain. Working through these issues will, in the end, tell us a great deal more about the nature of both birds and mammals.)

Amygdala. The amygdala too can be understood in part by noting its outputs. It sends massive connections to a small region called the hypothalamus, a set of regulatory structures that virtually runs the autonomic systems of your body. The hypothalamus operates your endocrine glands (testosterone, estrogen, growth hormones, adrena-line, thyroid hormone, and many others), and generates simple prim-itive behaviors that are appropriate to these hormones. As a particularly graphic instance, if you stimulate a righthand portion of an animal's hypothalamus, testosterone will be released into the bloodstream and the animal will immediately begin engaging in sex-ual behavior with whatever object happens to be near it. The amyg-dala largely rules the hypothalamus, and thus the amygdala is the forebrain regulator of these very basic behaviors. A closely related function of the amygdala is its evocation of strong emotional responses. Not only can it evoke primitive hard-wired behaviors via the hypothalamus, but it makes us feel emotions that correspond to those behaviors.

Enter again the irrepressible Professor Delgado. And this time, not bulls or monkeys but human patients. There were hospitals that were filled with schizophrenics, epileptics, and others who did not respond to any known treatments, and whose violent actions or seizures were deemed to represent an unacceptable danger either to others or to themselves. Delgado operated on dozens of such "untreatable" patients, implanting electrodes in their brains in hopes of providing them with a last hope of controlling their disorders. Stimulation in different regions in or near the amygdala could abruptly trigger extreme evocations of raw emotion. Brief electrical pulses could produce intense rage, or earnest affection, or despondent sadness, or almost any conceivable psychological state in between. The type of response was a result of the exact location of the pulse.

In one episode, Delgado and two colleagues at Harvard induced electrical stimulation in a calm patient, who immediately exhibited extreme rage and nearly injured one of the experimenters. (One of Delgado's colleagues was Frank Ervin, who had a medical student at the time who learned much of this material firsthand, and used it as inspiration for a novel he was trying to write on the side. That student, Michael Crichton, published that first novel, *The Terminal Man*, in 1972; it contained explicit scenes of rage-evoking brain stimulation. The book became a best-seller, and can still be recommended as an introduction to both the science and the potential dangers of these studies.)

These findings and many since that time have fueled speculation about whether emotional disturbances such as violent behavior or hypersexuality could reflect damage or dysfunction in the amygdala. For this reason, some potentially promising experimental drugs for treating anxiety and depression are explicitly designed to affect the circuitry of the amygdala.

Hippocampus. If any brain part could top the amygdala in notoriety, it would have to be the hippocampus, the third of the forebrain regions lying beneath the cortex. Where striatum and amygdala provide movement and emotion, the hippocampus is central to the encoding of memory. The chain of evidence began in the 1950s, quite by accident, when scientists discovered that the hippocampus and surrounding regions were often culprits in the generation of epileptic seizures. This led surgeons to treat intractable seizures by the expedient of taking out the entire offending region. That is, they removed the hippocampus and its surrounding regions, often including all or part of the amygdala and overlying cortical regions. (This substantial conglomerate of structures is sometimes referred to as the "medial temporal lobes," though this term of convenience may give the erroneous impression of uniformity to the very different internal circuits involved.) One of these surgical cases changed the history of neuroscience. A patient who we'll call Henry had increasingly serious seizures, from his teen years on, that did not respond to any treatments. By his late twenties he became so incapacitated that finally, in 1953, surgeons decided to remove a substantial region of his brain including his hippocampus. After the surgery, he still had seizures, though they were less incapacitating than before. Henry appeared remarkably normal for someone who had just lost a sizeable piece of brain. His speech, movement, sensory perception, and even his IQ appeared unscathed. But he, and other patients who had undergone similar surgery, were examined by a number of psychologists over the succeeding years, including Dr. Brenda Milner, who noted a marked memory deficit. Henry thought that it was still 1953, and had no memory of having had an operation. Moreover, he did not remember the doctor he had just been talking to, nor indeed did he remember *anything* that had occurred since before the operation. The shocking truth was that Henry apparently could no longer form new memories. He retained most of his past: he knew who he was, where he lived, recalled his high school experiences, and so on. But he seemed unable to add anything new to that memory store. His doctor could have a chat with him, leave the room, and then return, upon which Henry would deny that he had seen her before—not just on that day, but ever. To this day, he lives still frozen at that moment when his hippocampus and surrounding areas were removed. Neuroscientists have been drawn to the hippocampus, and its surrounding "medial temporal" structures, ever since.

It is noteworthy that the olfactory system sends its largest outputs to the hippocampus, and in turn the hippocampus, at least in rodents, receives its largest inputs from olfaction. The hippocampus clearly begins in small-brained mammals as a kind of higher-order processor of odor information. Indeed, rats normally remember odors extraordinarily well, but if their hippocampus is damaged, they have a difficult time learning new odors, a bit like Henry's difficulty after his operation.

Since memories from before hippocampal damage are intact, but new memories can't be created, it has often been hypothesized that the hippocampus is a temporary repository of memories which subsequently move onward to final, permanent, cortical storage sites. (This is sometimes likened to a hiatus in hippocampal purgatory before ascension into cortical memory heaven.) Indeed, if our memories were like books in a library, they might first go to a receiving dock, or a librarian's desk, before being permanently indexed and shelved. But our brains, of course, often use internal methods that don't always resemble the ways we perform everyday tasks. The hippocampus is likely, instead, encoding *contingencies* between events, the sequential occurrence of, say, a particular odor and a particular sight or sound. When the hippocampus detects a novel, unfamiliar contingency, it triggers a signal that alerts the cortex to store the new information. Without the hippocampus, these novelties won't be detected, and won't be stored.

Thalamo-cortical loops. The fourth and last target of the cortex is by no means the least: it is a connection to a large two-part system, consisting of another part of cortex (frontal cortex) and part of the thalamus (figure 6.1). This thalamo-cortical connection starts small, but as the brain grows large it becomes ever more important, until in big-brained mammals, thalamo-cortical circuits become the keystone of the brain. We will revisit these circuits in earnest in chapters 7 and 8.

The olfactory cortex generates connections to a particular thalamo-cortical loop, involving a specific region of thalamus and a specific region of cortex. The thalamic region targeted is named simply for its location, the dorso-medial nucleus or DMN. And the cortical region targeted by olfactory cortex is one of the most storied

pieces of the cortex: the frontal cortex. Frontal cortex, the most anterior part of the cortex, growing into the space between the nose and
the forehead, begins in small-brained mammals as a motor structure,
connected tightly to the striatum, which as we have seen is itself an
organizer of locomotion and other complex movements. As striatum
corrals the brainstem muscle-controlling structures, to produce
longer and more coordinated streams of motion, the frontal cortex in
turn examines those motions, and uses its learning mechanisms to
come to anticipate the outcomes of given moves. Since the frontal cor-
text can control the striatum, it can use its burgeoning predictive abilities to refine the selection of what movements to perform in what
situations. It thus becomes the beginning of a system for planning.

None of these systems acts in isolation, but rather each sends
certain specific types of messages to the others. In particular, both
the hippocampus and amygdala send their messages to the striatum, and the striatum in turn is linked, as we have seen, back to
other cortical and thalamo-cortical circuits. All the parts participate
in a unified architecture, as the different specialized components of
a car (spark plugs, transmission, linkage) are quite different when
they interact in an engine than when they operate by themselves.
Cortex and these subcortical systems are the five major engines of
the human brain. By observing how they interact, we gain our first
glimmering of coordinated intelligent behavior.

FROM CORTEX TO BEHAVIOR

Figure 6.2 illustrates the integrated connections among cortex and
its four targets. Striatum makes connections to frontal thalamo-
cortical loops, just like olfactory cortex does, thus creating even
larger loops: striatal-thalamo-cortical loops. And amygdala and
hippocampus in turn connect to the striatum, thus insinuating themselves into that large striatal-thalamo-cortical loop. As a result, every
odor experience ends up sending not one but multiple messages to
thalamo-cortical loops, one directly from the olfactory cortex,
another via striatum, and the rest via amygdala and hippocampus.

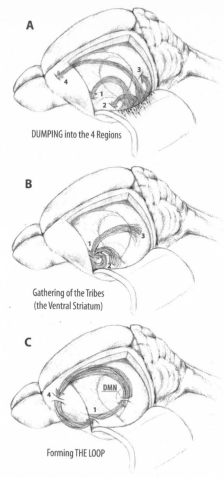

A

DUMPING into the 4 Regions

B

Gathering of the Tribes
(the Ventral Striatum)

C

Forming THE LOOP

Figure 6.2 (A) The olfactory cortex sends axons directly into its subcortical targets (see figure 6.1). (B) These targets all in turn project to the ventral striatum (1). (C) The ventral striatum sends part of its output to the thalamus (DMN), which then connects to the frontal cortex (4) which projects back to the ventral striatum, creating a connected loop. All targets of the olfactory cortex feed into one segment (striatum) of the loop, but only frontal cortex is positioned to regulate overall activity within it, which is why the frontal cortex can be thought of as the brain's chief executive.

For instance, frontal thalamo-cortical circuits receive information directly from olfactory cortex, and again indirectly from striatum. Combined, the frontal thalamo-cortical system has access to information both about the particular item recognized (e.g., the smell of food) and outcomes that have been associated with it (eating), and can make a primitive plan to approach the smell.

Meanwhile, the connection from olfactory cortex to amygdala elicits autonomic and emotional responses, e.g., thirst, satiation, lust, sleepiness. It sends these onward to the striatum, and thence to the frontal thalamo-cortical system. This can have striking effects. The hormonal satiation signals conveyed from amygdala and hypothalamus combine with a food's odor, altering the very sensation that we experience: the smell of food can seem very different before a meal versus after it. The amygdala can determine both the intensity and the valence, good or bad, of the experience that the frontal cortex will eventually perceive.

We saw that the connection from olfactory cortex to hippocampus triggered learning of contingencies among sequential events. The projection in turn from hippocampus to the striatum can be understood in the context of an example. Picture a hungry animal tracking an odor, through an environment rich with other odors, and sounds, and sights, all competing for attention. The world presents a constant stream of "blooming, buzzing confusion" to an organism; the hippocampus is a crucial player in sorting these myriad experiences into a semblance of order. When the animal has been in this environment before, the hippocampus has learned which experiences normally occur in this setting and which do not. While in pursuit of food, it can safely ignore extraneous sensations if they are recognized as familiar in this setting. But if it encounters an unfamiliar sound, sight, or smell, the hippocampus can trigger a "stop" signal in the striatum, to attend to and store the new information. Without the hippocampus, an animal will blow past a novel item, and new cortical memories won't be encoded.

With these components in mind, we can return to the frontal thalamo-cortical system. As described in figure 6.2, the outputs from striatum circle back to thalamus and thence to cortex, which in turn talks again to striatum. This forms a huge closed loop: the frontal cortex is receiving direct sensory information from cortex about the odor that is currently present, together with indirect information about the response just committed—whatever the animal just did. This information can then be used prospectively: it can influence what the animal will do next. Every past experience that the animal has had with these ingredients (e.g., odors and actions) enables it to learn the expected outcomes of various responses to different behaviors—and to use these to *choose* future actions. In past episodes of tracking or avoiding various odors of different kinds of food, animals, and environments, there were some with successful outcomes and others that were not; each event modifies the synaptic connections among links in the cortex and striatum. Thus, when returned to this setting, the odors and behaviors interact with striatal-thalamo-cortical circuits that have been shaped by hundreds or thousands of earlier experiences. These "experienced" circuits constitute a set of programs that can be used to switch behavior from one ongoing sequence to another.

Because these structures form a recurrent loop, the large frontal striatal-thalamo-cortical system can cycle for a long time—many, many seconds—and thereby produce extended sequences of behavior, and can "hold in mind" the current items that it is operating on. This "working memory" becomes one of the emergent tools available to the mammalian brain. This gives the frontal cortex the appearance of the brain's CEO, making plans and organizing the activities of disparate subdivisions to reach a goal. A first action will "prompt" the frontal cortical region to select a next action, which in turn prompts the cortex to select a further behavior, and so on, until, for instance, a sensed odor is located. Throughout this process, the animal will be engaged in apparently intelligent and seamless behavior.

From olfaction to other senses. All of this anatomy produces a surprisingly straightforward, and surprisingly complete, picture of how odors guide behavior. Cues are recognized by a primary cortex which then in parallel distributes signals to regions that initiate movement (striatum), intensify or weaken the movements (amygdala), detect anomalies during the search (hippocampus), associate the cue with objects (hippocampus again), and organize actions in appropriate behavioral sequences (frontal striatal-thalamo-cortical loops). The other sensory systems—vision, audition, and touch— follow the same basic pathways in connecting the outside world to useful responses.

NEOCORTEX

Two great splits occurred the long evolutionary history of the mammals. The first split gave us the monotremes: egg-laying creatures whose few descendants are still scraping out an existence in remote corners of Australia; the platypus and the echidna. The next split divided our ancestors among marsupials (kangaroos, wombats, opossums, koalas), who gave birth to live young, but kept them in a maternal pouch; and placentals, with live births and no pouch. Placental mammals are almost all the mammals we're familiar with: mice, dogs, bears, people. The earliest placentals—the stem mammals—had a lot in common with today's hedgehog. As the hagfish can be thought of as representative of the earliest vertebrates, so the hedgehog can serve as a model of early mammals.

The olfactory components of a hedgehog are the largest parts in its brain (see figure 6.3). The olfactory cortex of the hedgehog is about as large as the whole rest of the forebrain put together, and the amygdala and hippocampus, the targets that trigger emotions and contingent associations, are both prominent. The thalamus and

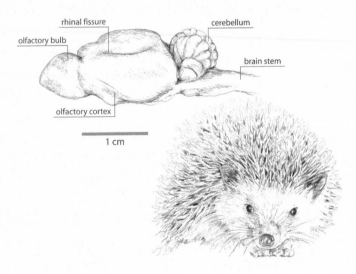

Figure 6.3 The hedgehog is representative of the earliest mammals. Its brain (top; also see figures 2.1 and 5.1) is dominated by the olfactory bulbs (left) and olfactory cortex, constituting most of the area below the 'rhinal fissure'. Other neocortical areas (above rhinal fissure) are much smaller by comparison, and the entire neocortex in the hedgehog is only marginally larger than the ancient cerebellum and brain stem.

frontal cortex are not so easily recognized as the previously discussed collection of structures, but they too are well developed in hedgehog brains. The visual, touch, and auditory areas appear to be pushed to the back of the cortex, leaving a broad area in the zone occupied by frontal cortex in other mammals. In all, the hedgehog forebrain is preoccupied with the olfactory system and its relationships, with the other sensory modalities having much smaller pieces of real estate.

Having been sequestered as nocturnal animals for so long, hiding from the dinosaurs, mammals' hearing and smell distance senses were developed, but their visual systems lagged behind. After the dinosaurs, the ensuing evolutionary changes included expansion and elaboration of the visual system. The midbrain areas below cortex, which had effectively handled both vision and hearing for eons of early vertebrate and even mammalian history, were soon dwarfed by the expanded cortical structures that became assigned to sight and sound.

The initial sensory cortical expansion was based on point-to-point organization, sending faithful representations of images and sounds forward into a neocortex that was, as we have indicated, set up with the random-access network designs representative of the olfactory cortex. Thus, the point-to-point world of the visual system becomes abandoned after just the first few connections of processing in the rest of the visual cortex. We have already seen that the same thing occurs in the olfactory system, where neat spatial organizations in the nose and in the olfactory bulb are replaced with scattered random-access representations in the olfactory cortex. As neocortex grew, it mimicked just this arrangement.

Once all the different sensory inputs—vision, hearing, touch— became encoded in the same random-access manner, then there were no further barriers to cross-modal representations. That is, the switch from specialized midbrain apparatus to cortical modes of processing allowed the brain for the first time to build multisensory unified representations of the external world. The result underlies the difference between the reptiles, largely lacking cross-modal representations, and the mammals, possessing them. Even the lowliest mammals appear in many ways cleverer, more intelligent and flexible, than most reptiles; it is the new mammalian brain organization, centered around the neocortex, that is responsible for mammals' more adaptive behavior.

Where then did the exquisite and specialized machinery found in the visual, auditory, and tactile regions of neocortex of modern mammals come from? We propose that these arose as secondary adaptations, features that sharpened the acuity of perception. There is a great deal more visual information than olfactory information to be processed by the brain, more than the retina and the first levels of visual processing can extract. We suggest that the visual and auditory cortex evolved as upper-stage sensory processors that supplemented the information-extraction of the initial brain stages.

A traditional view is that the point-to-point structures in our brains arose first, and the random access "association" areas arose later in evolution. Many of the small-brained mammals of today, such as rats, have a lot of point-to-point sensory cortex regions, but have relatively little random-access association zones. In contrast, today's larger-brained mammals tend to have more and more association areas. A natural hypothesis was that the original mammalian cortex was ratlike, dominated by sensory input with very little association cortex.

But, as we have pointed out, a rat is in no sense less evolved than a monkey. Rodents are, in fact, a more recent order than primates, having emerged only after the mammals invaded the post-dinosaur daytime world. Figure 6.4 summarizes both potential versions of the sequence of events that may have occurred as mammalian brains arrived at their combined point-to-point and random-access organizations. In the top of the figure, we see the hypothesis that the neocortex began with point-to-point designs, and later added association cortical regions. At bottom is the alternative hypothesis presented here: the neocortex began with an overall olfactory design, which humans retain and expand in our huge association cortical areas; the more highly specialized sensory regions for visual, auditory, and touch senses were initially small and were added to over evolutionary

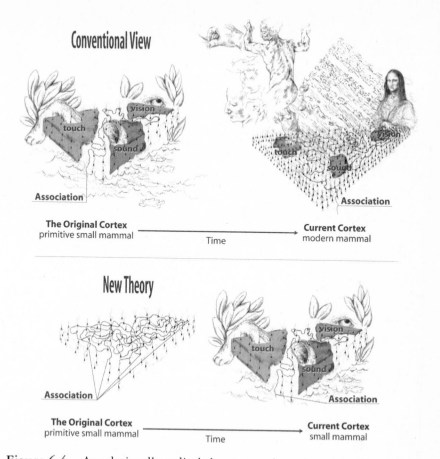

Figure 6.4 An admittedly radical theory as to how the great neocortex, by far the largest part of the human brain, came into being. (Top) The conventional view. The original mammalian cortex (left side) was dominated by distinct point-to-point zones making replicas of inputs from the visual (eye), auditory (ear), and touch (leg) receptors. Scattered between these were small, random-access association regions that generated combinations of these inputs. As the brain grew large over time, the point-to-point zones maintained their sizes but became surrounded by immense association regions, making it possible to assemble sound waves into symphonies and detailed images into paintings, and even to combine body maps with vision to create sculpture. (Bottom) The alternative. The neocortex in the first, small-brained mammals was based largely on the random-access olfactory cortex design. Touch, sounds, and visual cues were mixed together with few or no point-to-point replicas. With time, these latter systems were added and in small-brained mammals became dominant, but in big-brained creatures, the ancient association systems simply grew, and grew way out of proportion. The capacity for integration that is so characteristic of humans was, according to this argument, fully developed from the beginning of mammalian evolution.

time. Both modes of processing have critical uses, but as we will see it is the association areas that continued to grow explosively as the mammal brain expanded.

Another crucial observation further strengthens the hypothesis that the association areas may be more ancient than the sensory areas. Association areas connect heavily with the large subcortical areas described earlier: striatum, amygdala, and hippocampus, and the frontal thalamo-cortical systems, just as the ancient olfactory cortex does. Sensory areas do not make these connections. If point-to-point cortical areas came first, and association areas second, then the pathways connecting subcortical areas must have arisen later still; this ordering is extremely difficult to explain. It is far more likely that neocortex emerged using the ancient olfactory template, retaining its outputs to striatum, amygdala and hippocampus, and that the specialized point-to-point sensory areas were filled in later, and modified independently. Yet again, we find evolution re-using an ancient adaptation for a novel purpose.

CHAPTER 7

THE THINKING BRAIN

The human brain appears enormous, and indeed it is. The brain of an average successful mammal, say a lion, weighs less than a pound; a fraction of our three-pound brains. How did brains grow from their modest initial designs to our present-day human brains?

From the initial hedgehog-like brains of early mammals, which were still dominated by the olfactory system, mammalian brains grew the other sensory systems, especially hearing and vision. These new systems incorporated the same basic designs as the olfactory system, as described. Some present-day mammals have retained some of the intermediate features of these steps in brain evolution. A bushbaby is a member of the first subgroup of the primates; its brain looks like the illustration in figure 7.1 (left) in which we can see a well-developed olfactory system, but also the newly prominent other regions of neocortex, areas of vision and hearing. These correspond to the large expansions on the sides and in the back, growing to flop over the olfactory components and the cerebellum. This expansion of neocortex is even more pronounced in the marmoset (right side of figure 7.1), a new-world monkey. Each of these brain designs may be an illustration of what the brain was like at particular stages in our evolutionary development.

The layered, cloak-like structure of the cortex gives it a highly useful space-saving trick: it has developed into a crumpled shape to fit more

1 cm

1 cm

Figure 7.1   (Left) A bushbaby resembles an early primate; as with the hedgehog (figure 6.3), its brain has very large olfactory bulbs but now added to a neocortex that is so large it exhibits the cortical folding characteristic of the largest mammalian brains. (Right) The marmoset, a new-world monkey, resembles the ancestor of the monkeys and apes. Its olfactory bulbs are greatly reduced and the neocortex now completely dominates the brain.

compactly in the skull, like a rug being shoved into a too-small room (as we illustrated in figure 5.1 in chapter 5). In big-brained mammals, the cortex takes on an elaborately folded appearance. In the bushbaby, we see the first appearance of these folds and wrinkles, as the cortex grows too large for the skull. These wrinkles are more pronounced in the marmoset; these and other monkeys evolved about 35 million years ago, and their brains continue to follow this trend of an ever-larger cortex.

As brains grow, most of the added structure is not in the sensory areas, but in the association cortex. These disproportionate rates of growth, as we will see, determine the overall percentage of the brain dedicated to the senses versus that dedicated to association. A relatively small-brained mammal like a bushbaby has about 20 percent of its brain taken up by visual cortex. But as the brain grows with evolution, visual cortex grows at a slower rate than association cortex. So in bigger-brained mammals, association cortex catches up and passes sensory areas. This trend keeps up through huge-brained humans; we have only small percentages of our neocortex dedicated to the senses, and all the rest is association cortex as illustrated in Figure 7.2.

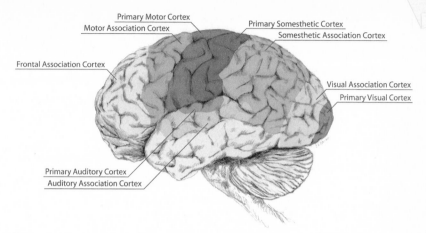

Figure 7.2 The human brain and its vast tracts of association cortex. The few dark areas are the primary point-to-point zones creating replicas of real world stimuli. Surrounding them are large association regions (light gray) specialized for combining images into artwork, sounds into music, and muscle movements into actions. And beyond these are still other broad association territories (not shaded) that go far beyond the sensory and motor worlds into the realm of thought.

Figure 7.3 illustrates the results. In the upper left are outlines of the brains of small-, medium-, and large-brained mammals: a rat, a monkey, and a human, showing their different sizes.

When we scale these up to be roughly the same size, we see the differences in organization that arise from the different rates of growth in point-to-point sensory areas versus random-access association regions.

In the rat, much of the brain is taken up by vision, hearing, olfactory, and combined touch/motor zones. We also see the large olfactory bulb at the front of the brain (left) and ancient cerebellum at the back (right); precious little is occupied by association zones.

In the monkey, the association zones have grown much more than the sensory areas. The monkey's sensory areas are somewhat larger than those of the opossum, but they take a much smaller percentage of the monkey's larger brain.

In the human, the sensory zones are larger, in absolute terms, than in the monkey, but barely. However, they occupy only a small fraction of the greatly enlarged neocortex. All the rest is association cortex (compare with Figure 7.2).

During these expansions, there is no hint of evolutionary pressures for more association cortex. Instead, the human brain is

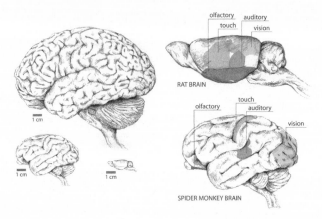

Figure 7.3 A summary of how the expansion of brain disproportionately expands association regions relative to sensory processing. The left panel shows brain outlines of a rat, a monkey, and a human. Most of the rat cortex (top right) is dedicated to hearing (medium gray), touch (lighter gray), and vision (lightest gray), as well as olfaction (darkest gray; note the large olfactory bulb and adjacent olfactory cortex). In the monkey brain (bottom right), these areas are larger in absolute size but occupy far smaller percentages of the larger brain. The trend toward different proportions is continued and amplified in the human brain.

simply a greatly enlarged version of the original mammal brain. We revisit this point in great detail in chapter 11.

In all these brains, the organizational layout for the senses stays the same. Just as we saw in olfaction, initial point-to-point inputs give way to downstream random-access circuits. For instance, in vision, cells in the eye send their axons to a group of neurons in the thalamus, which in turn send their axons to the primary visual cortex, and throughout these initial stages, the image on the retina is projected, point-to-point, all the way to the cortex. The same process holds for hearing and touch. Those primary sensory areas then send connections into association areas that lose their point-to-point organization and acquire random-access organization. From there, connections go to the striatum, the amygdala, and the frontal thalamo-cortical system, according to the same overall plan we saw in chapter 6, allowing the associational cortex to generate movements, trigger emotional responses, and engage the planning system.

Critically, the associational cortices also project to each other, and the larger the brain, the more of these connections there are. As noted earlier, these cortico-cortical connections make sense only because all of these areas are using the same system for laying down their representations. The memory of a friendly face is stored in one association area as a pattern of active neurons, a set of addresses, while the memory of the person's voice is simply another collection of addresses in another associational area. All that's needed is a "downstream" association area that can be reached by both the face and voice representations. Those downstream memories will constitute a more abstract memory: a representation of the link between that face and that voice.

The system is thus hierarchical—lower, simpler levels send messages to higher, more abstract levels, and receive feedback in return. Associational regions that are dominated by inputs from vision tend to connect to each other, but also connect to areas dominated by auditory input, as well as to areas whose input is not dominated either by vision or sound alone. Go downstream far enough and there will be a region where information on face and voice can be combined into a single brain code.

EXTENDING THINKING OVER TIME

We have so far focused on association cortex in the back part of the cortex. This houses the pathways that collect input from sensory areas—vision, hearing, touch—and integrate them. There remains a vast territory, perhaps a quarter of the entire cortex, at the front of the brain, under your forehead, labeled the frontal association fields. That forehead tells the story: only humans have them. In most animals, the head sweeps aerodynamically back from the nose, because they have much smaller frontal fields. The larger the brain along the evolutionary ladder, the disproportionately bigger the frontal fields get. In primates, and most noticeably in us, those frontal areas become enormous.

Outputs from the frontal cortex provide more clues to its nature. There are three primary output paths. The first goes to the motor cortex, which contains a point-to-point map of the body's muscles. You want to move your left foot? An output signal moves from the frontal cortex to the premotor and motor cortex, triggering brainstem and muscle activity. The second pathway from frontal cortex connects to the striatum, which in turn projects right back to thalamo-cortical circuits, thereby creating the closed loop we described in chapter 6: cortex to striatum to thalamus to cortex. The third and final pathways are two-way connections from frontal cortex to the sensory association cortical areas and back again. Taken together, these connection pathways explain three primary roles of frontal cortex: planning movements, timing them, and coordinating internal thought patterns.

Motor movements are thus driven by abstract units, which we can think of as providing planning: an abstract time-and-motion map of what movements to carry out. A baseball pitcher has to arrange his

Figure 7.4   Successive brain regions are engaged as we move from plan to execution. The frontal cortical regions select a pitch (slider, change-up, etc.) from a repertoire of stored (learned) possibilities and initiate appropriate movements by activating the motor cortex and its direct axon projections down to 'motor' neurons in the spinal cord. These latter cells cause the muscles to contract. As the messages move from the cortex downward, they send side branches to the cerebellum and brain stem (placards in the drawing) to insure that all parts of the body work in harmony. They also engage the striatum and its 'loop' with the frontal cortex; the loop stretches the time span of frontal activity, so the entire sequential program can run its full course. Finally, inputs to the frontal areas from sensory association cortex provide constant updating of where the body is relative to the program's targets.

body movements so that the ball will leave his hand and then a half-second later curve across the plate. All the moves have to be carefully triggered: move this leg muscle, then this left arm muscle, and then release at this moment in the sequence (figure 7.4). The projections from frontal cortex to motor cortex provide that hierarchy of abstraction from movements to planning.

Our physical movements take place over time spans that can take seconds, but neurons operate in fractions of a second. The connections from frontal cortex to striatum are involved in this process. Striatal neurons can change their voltages for long periods of time; more than enough for the production of a sequence of behavioral actions. These same striatal neurons also receive other inputs from a particular type of neurotransmitter, dopamine, that can produce seconds-long effects. Combined, these features make the striatum well suited to stretch activity out across time. And since the projections from striatum back to frontal cortex create a closed loop, the striatum can "inform" the cortex of the time spans needed to produce serial sequences of behavior.

The third and final set of connections, from frontal cortex to association regions in sensory cortex, connect a planning region with areas that we described as responsible for internal feats such as combining information about a face and a voice. These provide the final ingredient needed to coordinate complex behavioral sequences. Unified information about our sensory perceptions allow us to walk smoothly, to learn to pronounce words to sound the way we want them to, and to parallel park our cars. In some exceptional cases, it allows a pitcher to precisely adjust his delivery of the ball to generate a strike.

The system does more than just organize actions, though; it also enables the organization of thoughts. Since the frontal cortex can assemble long temporal sequences, we can construct and reconstruct memories of past episodes. The system lends itself to creative use: frontal cortex essentially has its pick of all the vast amounts of sensory material stored throughout association cortical regions. This ability plays out most in brains with the largest frontal cortex: human brains have vast frontal regions, whereas these are of modest size in most other mammals. In some brains, this no doubt will result in new and unexpected combinations, perhaps even novel assemblages that bear the mark of genius.

There is room for error in these mechanisms. How does the frontal cortex know that the material it organizes from storage actually relates to anything in the real world? Some schizophrenics suffer from hallucinations in which they hear voices that are as real to them as those coming from real world speakers. The disease involves the frontal cortex and its dopamine-modulated striatum loop; possibly an instance in which the novel assembly of internal facts goes awry. Perhaps regular reality checks are needed in all of us, constraining the creative construction process taking place internally.

THE CORTEX TAKES CHARGE

In the underappreciated 1950's movie *Forbidden Planet*, a classic mad scientist, Morbius, discovers that a race of very advanced beings on the eponymous planet were completing a colossal machine to convert thought into matter just before they suddenly and utterly vanished. A researcher from a visiting spaceship uses a piece of technology left over from these godlike creatures to boost his intelligence far beyond the normal range. Just before dying (the fate of all who steal from the gods), and with his last breath, he gives a terrifying clue about the advanced beings' disappearance: "Monsters, Morbius, monsters from the Id!" As it happened, the long ago race had actually finished their machine and that night, when asleep, the ancient part of the brain, the home of Freud's Id, took control and made material all the hates and desires suppressed by a civilized veneer. They destroyed themselves in a single night.

This story nicely captures a popular view of the brain—that lurking beneath the cortex lies a more primal, sinister, and suppressed aspect, with reptilian-like impulses. As we've seen, the major divisions of the brain are found all the way back in the hagfish; so the beast within us is perhaps more fish than reptile. But whatever the "flavor" it may possess, that lower brain does indeed compete with the cortex.

There are a number of "modulation" systems that travel from the lower brain to cortex; we can think of them as "dials" that can be turned to adjust the behavior of higher brain systems. We briefly discussed dopamine, which can reward or punish cortical and striatal behaviors, thereby influencing future choices. There are other modulators, such as norepinephrine and serotonin, each with their unique and powerful effects on the cortex. All of these ancient brain regions and modulatory projections, which we inherit directly from the earliest vertebrates, are activated and deactivated by basic biological variables such as body temperature, hormone levels, time of day, hunger, and so on. Higher regions, receiving these messages, are coerced to change their behavior. The amygdala, for instance, which we showed generates emotion-related behaviors, is strongly affected by inputs from the ancient autonomous regulatory systems. The less cortex in a brain, the more it is dominated by these lower mechanisms. As the cortex expands, the disproportionately increased association areas provide ever-greater influence over structures like the amygdala. The bigger the cortex, the more it wrests control from the whims of the ancient projections.

The battle is never completely won. The touch of the ancient modulators is felt in mood, quality of sleep, the surreal world of dreams, and the nervousness triggered by flashing red lights. Anti-anxiety treatments such as Prozac, Zoloft, and Paxil are aimed precisely at one of these modulators, the serotonin system. It is even possible that the enormous human association cortex can be trained to supernormal levels of control over the serotonin, dopamine, and other systems. Perhaps here is an explanation for the lucky or talented few who carry on in the face of tremendous stress, or for the inner calm that certain esoteric practices are said to produce.

The lower brain, hiding under the cortical mantle, is the material from our vertebrate ancestors that gave rise to reptiles, birds, and mammals alike: the ur-vertebrate. Monsters from the Id.

These lower circuits are always there, under the surface, generating hard-wired responses to environmental signals; collections of ancient neurons trying to assert the needs of the body. And higher association zones are in constant struggle with them.

Success in controlling the lower brain depends on the relative size of the cortex. Humans, with their vast association regions, have more brain intervening between these lower regions and their behavior than do any other animal. Perhaps most important in separating man from the beast is the machinery described above for thinking and imaging. Chimps have been shown to make fairly elaborate plans, and gorillas at times seem pensive. But the human brain's association regions, the dense connections between them, and our huge frontal area with its striatal loop, are many times larger in humans than in apes.

CHAPTER 8

THE TOOLS OF THOUGHT

FEEDBACK AND HIERARCHIES
OF CORTICAL CIRCUITS

We have seen that the thalamo-cortical circuits of the brain work intrinsically in concert with the major subcortical systems (striatum, amygdala, hippocampus), but also with other cortical systems. As brain sizes increase over evolutionary time, cortical systems increasingly connect with each other, performing more and more operations beyond those that a small brain can carry out.

We earlier raised a key question: if the new mammalian neocortical systems retain their structure, and simply grow to constitute more and more of the brain, how do these same repeated circuits come to carry out new and different operations? How can a quantitative change—simply making more of the same cortex—generate qualitative differences in different animals? How do dogs get smarter than mice, and monkeys smarter than dogs; and in particular, how could the unique faculties of humans—planning, reasoning, and language—arise from having just more of the same brain?

We begin with the seemingly simple response of cortical circuits to a visual image. A picture of a flower activates cells in the retina, in a spatial fashion that mirrors the pixels in the flower image. The incoming image selectively activates neurons in the retina. Those retinal neurons' axons project in point-to-point fashion into the thalamo-cortical system, which then projects to downstream areas that have random-access connectivity, as we have described in the last two chapters.

We now introduce the final crucial point about cortical circuits: they generate projections both forward, to downstream areas, and backward, back to their inputs. This occurs both in thalamo-cortical and in cortico-cortical circuits: messages are sent in both directions, implying that the initial processing of an input—a sight or a sound—becomes altered by the downstream conception of what that sight or sound may be. Our "higher level" processing actually modifies our initial processing. Perception is not pure and direct; it is affected by our learned "expectations." So our prior experience with flowers—what they look and smell like, where they occur, when we've seen them before—all can quite literally affect the way we perceive the flower today.

One effect of these projections is the re-creation of sensation. The random-access abstract representations in sensory association areas are capable of reactivating the point-to-point maps in sensory cortical areas, re-creating realistic images of the environment. A painter can envision a picture before it is on the canvas; Beethoven, almost deaf, could hear the Ninth Symphony in his mind, before it existed in sounds.

There are further implications of these brain circuits, and more surprising ones. As we discussed in chapter 2, since the tools to study cortex directly are very limited, many scientists construct computer simulations to explore brain circuit operation. Sufficient computational power is now available to build models of neurons, each imbued with biological properties found in real neurons, and then to string together thousands to millions of them, following the circuit designs dictated by anatomy.

Models of this kind are sometimes sufficiently rich in detail to do things that were never anticipated by their creators. We will see an instance of this here. We will describe the steps carried out by a computational cortical simulation, and see the surprising results of its operation. Even an apparently simple response to a visual image is not straightforward: "simple" recognition involves multiple hidden steps.

Step 1: Initial activation

A particular image of, say, a flower, will activate some pattern of cortical neurons in the computational model of the visual system, corresponding to the features in the image. A different flower will activate another cortical pattern—but by definition, shared features between flowers A and B are likely to activate overlapping cortical cells, since those target cells selectively respond to the occurrence of those features.

In the accompanying schematic figure (8.1), some neurons in a target population are activated by images of different flowers. Each flower activates different axons (horizontal lines), with thicker lines denoting those that are activated. Each axon sends its message, if any, to neurons that it makes synaptic contact with. For instance, following the topmost axon in the figure, going left to right, we can see that it makes contact with the first neuron, but not the second or third, and then contacts the fourth, etc. The three flowers, though different, all share some features: petals, circular arrangement, some colors. In this illustration of the computational model, those shared features are presumed to be transmitted through the top two axons for each flower; those axons are shown as thicker than the others. We can see that all three flowers, despite their differences, activate those two axons. The different features of each flower are transmitted in each case down a different axon: the fourth axon down for the rose, and the fifth and sixth axons down for the next two flowers. In the computational model, the field of target cortical neurons receives input transmitted from the active axons in response to one of the flower pictures. A neuron that receives too little input will remain inert (dark), whereas neurons that receive the most activity will themselves become active (bright). In the model, neighboring neurons inhibit each other, thereby "competing" for activation: if two neighboring neurons are triggered by an input pattern, typically only the most strongly activated neuron will actually respond.

In the figure we can see that, from left to right, neurons 1, 4, 9 and 10 are activated in response to the rose; neurons 1, 4, 9 and 11 are activated in response to the daisy, and neurons 3, 4, and 9 are activated in response to the violet.

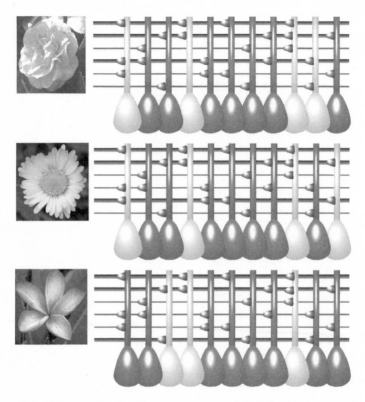

Figure 8.1 Eleven simulated neurons respond (brighten) to a rose (top), to a daisy (middle), and to a violet (bottom). Neurons respond if they receive sufficient activity from the signal arriving at the eye (left), via connecting synapses, interspersed through the field of dendrites. Initial responses to different flowers overlap, but are not identical.

Step 2: Learning

Each episode is learned in the activated brain cells; i.e., each time a feature is seen, its synaptic connections are strengthened in the computational model. The next figure (8.2) illustrates these synapses before any learning (left, same as 8.1 above) and after multiple episodes of experience with many flowers (right).

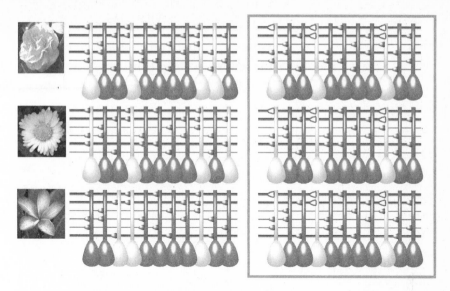

Figure 8.2 Responses of these same eleven neurons before (left) and after (right) learning. Synapses that have been activated the most will strengthen (white synapses in the right half of the figure). These stronger synapses can overcome slight differences among images (flowers), causing the same neurons to respond to any flower (right side of figure).

With repeated experience, the synapses in the model that are most often activated become stronger. Axons that are shared, i.e., that participate in more than one flower image, naturally tend to be activated more often than those that occur only in a few instances. The righthand figures show shared (and thus differentially strengthened) synapses in white. As these become stronger activators of their target neurons, those neurons become increasingly likely responders to any flower, of any kind. Moreover, since cells "compete" with their neighbors, inhibiting those that respond less robustly, the most strongly responding cells increasingly become the only responders. We can see that, after learning, the strengthened synaptic connections cause neurons 1, 4, and 9 to respond to any flower, whether rose, daisy, or violet.

The effect of learning, then, is to prevent these target cells in the model from differentiating among slightly different inputs.

At first glance, this is a counterintuitive outcome. Surely learning makes our responses better; smarter; more differentiating. Yet this finding in the computational model suggests that learning renders us *less* capable of making fine distinctions. The dilemma is immediately resolved in two ways.

First, note that there is some demonstrable value in eliminating fine distinctions. Eleven slightly different views of a rose are still that rose. If every different view triggered a different pattern of cortical cell activity, each view would correspond to a different mental object. Indeed, if every separate view was registered as an entirely different percept, we'd be overwhelmed by the details of the sensory world, unable to recognize the patterns of similarity that recur. Clustering gives rise to internal organization of our percepts; it enables generalization from individual flowers to the category of all flowers.

Second, recall that all of the processing just described proceeded via a forward-directed circuit, from thalamus forward to the cortex. As we've just described, there are also backward-directed circuits, flowing the other direction: feedback from "higher" back to "lower" cortical areas, and feedback from cortex back down to thalamus. These feedback pathways now play a role.

Step 3: Feedback

Once a "category" response has been elicited from cortex, feedback signals are sent back to the input structure, the thalamus. This feedback projects to thalamic inhibitory cells, and selectively suppresses part of the input—just the portion that corresponds to the cortical response, which is the shared or category response, shared for all flowers.

The flower is still out there, and the cycle begins again: the eye sends signals to the thalamus, which will send signals up to cortex— but inhibition is long-lasting, and part of the thalamus has just been inhibited by the cortical feedback. So only part of the input makes it from the thalamus to the cortex: the part that is "left over" after the operation of inhibiting or subtracting the shared "category" components of flowers. That leftover, that remainder, contains features not shared by all flowers—features that instead are unique to this particular flower, or at most to some subset of flowers.

That "remainder" signal now flows up to cortex—and cortex will now respond, just a few tens of milliseconds after its initial "category" response. This second response in the computational model is triggered only by the "remainder" inputs, so it elicits a completely different pattern of cortical target cells than the first input did. Over many episodes, learning selectively strengthens cortical responses to features shared by all flowers, and by the same token, learning strengthens subsequent responses to any features shared by particular subsets of flowers. Feedback inhibitory subtraction has the same effect on third responders, and fourth—and the inhibitory signal finally fades after about four or five such responses (corresponding to up to a thousand milliseconds, or one second).

What does this mean? The cortex seeing a single flower, but emitting a series of four to five quite different responses over time?

Analysis shows that, as we described, the first responders will be the same to any flower; the second responders will be shared for all roses, or shared for all violets, or for all daisies. Later responders will correspond to even smaller subgroups such as white versus yellow daisies, and eventually, cortical responses will be selective to a category that may contain a single particular rose or daisy.

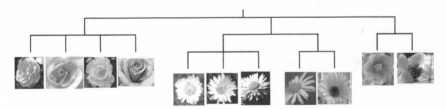

Figure 8.3 Synaptic change causes responding neurons to respond identically to similar inputs, and thus has the effect of organizing percepts (flower images in this case) into groups and sub-groups. Computational models of thalamo-cortical loops iteratively "read out" first membership in a group (flower), then sub-group (large vs small petals), then sub-sub-group (daisies). Evidence exists that human brains operate in this way.

What has happened is automatic, and occurs with no overt training, but rather simply by experience—and the result is astounding. Having seen instances of roses and daisies, this thalamo-cortical system slowly, over repeated exposures, stores its memories of flowers in such a way that that memory acquires internal organization, sorting flowers into categories (see figure 8.3). From then on, when any particular item is seen, the thalamo-cortical system produces not one "recognition" response but rather a series of responses, "reading out" first that the object is a flower, then that it is a daisy, then that it is a yellow daisy—traversing down the hierarchy from category, to subcategory, to individual.

What these brain simulations suggest is that recognition is not a unitary thing: we recognize over time, with discrete ticks of the clock producing additional information about the object being viewed. Multiple "glances" tell you a succession of different things: category, subcategory, and on to individual objects. The whole process unfolds in the blink of an eye; within a fraction of a second. These low-level thalamo-cortical and cortico-cortical circuits carry out unexpectedly complex sensory processing, and memory organization, and interaction between memory and vision, all in a few moments of perception.

When these computational models first were constructed, there was precious little evidence to suggest one way or another whether such behavior might actually occur in the brain. Since that time, studies have begun to appear indicating that something very much like this may indeed be going on in your brain: no matter what you see, you first recognize only its category, and only later recognize it as an individual. The implication is that recognition is not as we thought: recognition occurs only as a special case of categorization, and subcategorization. It is not a separate, or separable, brain operation, but an integral part of the process of category recognition. In the next section, additional models of cortex will elicit further findings that are similarly counterintuitive.

SEQUENCES

In addition to the categorization responses just described, cortical circuits string these category responses together into mental "sequences." This process has been repeatedly found by many researchers, and published in the scientific literature over many years.

If one petal of a flower is seen, and then another petal, cortical circuits link these together into a sequence that identifies a relation between the two petals. Such a sequential relation might be, "move to the right one petal-width and down two petal-widths," as though describing the eye movements themselves that would have to be carried out to traverse from one petal to another.

Recall that these sequences are themselves sequences of categories; rather than sequences of a specific flower petal, they will tend to describe sequences of petals in general. These linked structures, sequences of categories, form the elemental memories that your brain creates. We have hypothesized that all memories are constructed from these parts. We now investigate what happens when a complex scene is reconstructed from sequences and categories by these brain circuits.

WHAT ONE BRAIN AREA TELLS
ANOTHER BRAIN AREA

There are perhaps 100 billion neurons in your brain, each of which may, at any moment, send a signal via a brief, tiny pulse of electricity to other neurons, via roughly 100 trillion connections or synapses. Each neuron can, in turn, re-route the message to still other target neurons. The spreading activity, coursing through the pathways of the brain, constitutes the message. That activity is the substance of thought. The natural question is how such activity patterns can underlie thinking, and that question will be addressed repeatedly through the rest of the book.

First, we ask what tools there are to observe the activity; what methods we have to enable us to watch the brain in action. In a system that is constructed by nature, rather than by engineers, we have no specification sheet or instructions telling us what parts there are, let alone what they do; we can only rely on experiments to tease out the nature and operation of these mechanisms.

Yet we can only do certain *kinds* of experiments. If we could observe all the chemical and electrical activity of each individual brain cell or neuron, while we put a brain through its paces—recognizing objects, learning, remembering, talking—then we could begin to pile up masses of data that we might then sift through to understand how brains operate. Even then, there would be an enormous flood of data, and the task of interpreting those data would be daunting. But that's the ideal situation: having data about the complex activity occurring in neurons during behavior would make this daunting job far easier than it actually is. In fact, we can collect almost no data of this kind.

Start with the experimental machines. The best devices we have for observing brain activity are far from what we would wish for. Experts in the field of neuroscience are in ongoing debate about the nature of brain signals, and we will give just a bare introduction.

The electrical activity that occurs in the brain can be measured via electroencephalograms (EEGs) or magnetoencephalograms (MEGs). These systems accurately sense the rapid time course of the messages sent from one brain area to another—but they can barely tell us where those signals originate or arrive, giving coordinates across broad brain areas, rather than individual neurons. In contrast, fMRI ("functional magnetic resonance imaging") yields increasingly closer and closer portraits of brain regions (though still at best constituting clumps of hundreds of thousands of neurons!), but its measurements are of activity over the course of seconds, incredibly slow compared to the electrical activity that can occur in one thousandth of a second. There are variants and compromises among these methods (PET scans, CAT, NIRS), but at best, we still grossly trade off accurate timing for accurate locations, generating maps of brain activity that are either blurred across the brain's surface, or blurred across the time of the message. It is as though we could retrieve satellite recordings of telephone conversations, but either all the conversations of each entire country were summed together (as in EEG) or the individual recordings of a modest-sized city could be distinguished, but only the average volume was captured, not actual words or sentences (fMRI).

WHAT'S IN AN IMAGE?

The pictures produced by these neuroimaging methods have become familiar: at first glance, they appear simply to be brains with bright splotches superimposed (see figure 8.4).

The way these images are produced is very involved, and is not at all a simple snapshot of actual activity in the brain. They're the

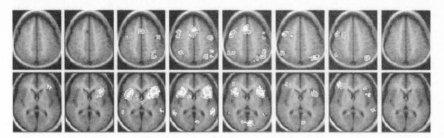

Figure 8.4 Images created from functional magnetic resonance imaging (fMRI) of a human brain. The bright spots are areas where more blood oxygenation occurred, implying that *relatively more* cortical activity occurred in the corresponding brain regions. The remainder of the cortex is also active, just not quite to the same degree as the indicated zones.

product of extensive interpretation, reflecting very slight but reliable differences in the amount of activity in a brain area during a particular behavior. Simple instances are easy to interpret: when we look at an object, areas of our brain dedicated to vision are more active; when we listen to sounds, areas dedicated to hearing are more active. But far more subtle differences also arise: slightly different patterns of activity arise when you look at a car, a house, or a face. One way of thinking about it has been to tentatively assign functional names to differentially responding areas. That is, if an area responds slightly more to images of desks than to any other images, whether houses, hammers or horses, it might be termed a "desk area." Or, more generically, if areas respond more to where an object is in a picture, than to what the object is, it might be termed a "where" area (distinguished from "what" areas).

Let's examine these responses more closely. If a "desk area" responds differentially to desks, how does it do so? A most likely answer arises from the paths traversed through cortex, from the eyes all the way in to the purported "desk" area (and other areas).

PUTTING IT TOGETHER: FROM GENERALISTS TO SPECIALISTS

When we see neural activation images, it is natural to think that we are seeing "the" areas that are "performing" a function. Thus we may hear the suggestion that, for instance, brain area X is "the" structure that "recognizes" a musical tune, or area Y recognizes your grandmother's face. By analogy, the wheels are the part of a car that move it along the road—but the wheels are rotated via differentials, which are activated by a drive shaft, which is operated by a transmission, which is powered by an engine, and so forth. These systems are designed to be distinct modules, which can be built and tested by people, in factories. Biological systems can be at least as complex, and can operate with at least as much interaction among components. Thus sounds enter your ear, activate your tympanic membrane, vibrating your cochlea, which sends coded electrical signals to an ancient structure in your brainstem, thence to a select group of neurons in the thalamus, and then to a series of successive cortical regions. Does one of those regions "recognize" the tune (or bird song or human voice)? Are cortical areas modularly allocated to functions that we have convenient names for, such as the "voice" region, the "bird song" region, the "music" region?

Again, controversy. Opinions range (and rage) between extremes. At one extreme, we have the dismissively termed "grandmother cells," i.e., those that respond always and only when your grandmother is present, in any lighting and in any attire. These seem in some ways like a caricature, yet some cells like this may exist in the brain. At the other extreme, we have entirely "distributed" representations, the antithesis of grandmother cells, in which it takes large populations of cooperating neurons to represent any complex entity such as a particular person. In these latter distributed representations, any individual cell recognizes only specific, low-level features, so the presence of your grandmother is signaled by the co-occurrence of all her individual grandmotherly features. But closer examination reveals that these positions are less distinct than they initially may seem.

MEMORY CONSTRUCTION

How are the basic ingredients combined? How does perception of an edge, or a line, lead to perception of a car, or a grandmother?

Earlier in this chapter, we saw the basics of thalamo-cortical circuits. "Front-line" neurons respond to direct physical stimuli, and after these initial responders, all other neurons respond to simple categories of similar signals, and to sequences of these categories. Simple features recognized by front-line neurons include brief line segments, oriented at angles from vertical to horizontal (see figure 8.5):

Figure 8.5 Initial neurons in visual cortex respond to simple features such as differently oriented lines and edges.

These in turn combine their messages to selectively activate a second rank of downstream neurons, responsive only to specific combinations of the simple first-rank neurons. Thus if there are simple front-line cells A and B that respond to a horizontal and a vertical line segment, respectively, then a second-rank cell C might respond only when both a horizontal and vertical line are present, as in the sight of an "L," a "T," or "V," or a "+" (see figure 8.6).

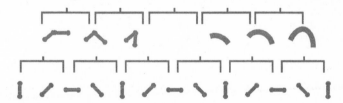

Figure 8.6 Neurons further downstream respond to combinations of simple initial features, such as oriented edges assembled into angles and shapes.

By a cascade of increasing selectivity, neurons further downstream might respond just to a box, or to other simple patterns; curves, circles, angles. If early cells respond to the simple circles of eyes, horizontal line of mouth, and so on, successive combinations might respond only when those features are in the positions they assume in faces. And if some cells are activated or excited, whereas others are selectively suppressed, by particular arrangements of inputs, it is possible to carve highly selective responses in some far-downstream target neuron. Such a neuron might selectively respond to the particular features of a particular face—and, conversely, any particular face might "recruit" or activate anywhere from a small to a large collection of such cells (see figure 8.7).

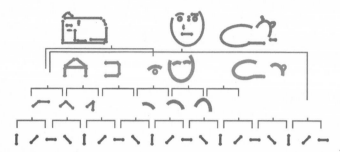

Figure 8.7   Progressing further downstream through cortex, groups of neurons respond selectively to increasingly complex combinations of features, such as those that occur in houses, or faces, or animals. These sequences of categories form internal "grammars" selectively responding to different percepts.

Evidence exists for both positions, for grandmother-like cells, and for fully distributed representations. And that same evidence also supports more complex but possibly correct intermediate positions, such as what we've just illustrated: that the number of neurons activated depends on the image shown. Neurons exist that are exquisitely tuned to particular complex images, and a given image may activate different-sized collections of cells, depending on the viewer's prior exposure to this image, and to other images that either share its features or have been categorized or associated with it. The association cortices proceed hierarchically, building ever more complex representations condensed into further and further downstream cortical regions. Connection paths through the brain take a range of directions, branching from initial generic features, to certain objects, to subgroups of objects. As we progress inward, further along the process, following various brain paths, we reach regions that are increasingly "specialized" for the particular assemblages of inputs that they happen to receive. All these brain paths are traversed in parallel with each other; the ones that respond to a given sight or sound are the ones that we perceive as registering recognition of a memory.

BUILDING HIGH-LEVEL COGNITION

As we have seen, the larger the brain, the more cortical association areas. Thus, deeper and more complex hierarchies are constructed by learning. Successive cortico-cortical areas build up from simple features to faces and houses, and with more cortex, more specialists are constructed. Just as faces and houses are built via relations among constituent features (eyes, nose, mouth; walls, windows, roof), these far-downstream cortical areas begin to build specialists that register increasingly elaborate relations among objects and actions. This territory of deep cortical areas has traditionally been hard to label in terms of function. Scientists have easily labeled the first stages of cortex for their sensory and motor functions: visual cortex (V1), auditory cortex (A1), motor cortex (M1). But down-stream areas have been lumped together as "association" cortex, with individual names typically afforded only by numbers (V2, V3, A2, A3, . . .), or by their relative locations in the folded mass of brain surface: medial temporal cortex; posterior parietal cortex; angular gyrus; dorsolateral prefrontal cortex . . . .

For each such cortical area, we can study two crucial aspects of its nature: first, what other areas it connects to, receives information from, and sends information to; and second, what circumstances tend to selectively activate it. Connectivity clearly defines some areas: if a cortical region, such as "V2," receives its primary input directly from early visual areas (V1), then that receiving area is probably responsible for learning slightly more complex visual constructs, and so on with successively deeper areas along this set of visual brain pathways. These pathways rapidly fan out, and begin to merge and to fan out again, creating a network of downstream areas whose function can't be determined from its connectivity alone. We have pointed out that neuroimaging techniques, from EEG to fMRI, have some ability to scan our brain while we are engaged in simple tasks. Thus, these methods have a rudimentary ability to highlight which brain areas are more active than others during different tasks.

Neuroimaging has indeed begun to separate some areas from others, enabling tentative identification of particular brain regions with particular mental operations. Some of the resulting operations are difficult to capture in everyday language; a given brain area may be computing an unnoticed function that underlies a number of our abilities. An example is the task we saw earlier, of building hierarchies of object representations; until careful experiments were carried out, it was far from obvious that people were recognizing not only faces and houses, but also were recognizing un-named partial assemblies that participated in many different images. Those barely-conscious partial constructs underlie our ability to rapidly recognize complex scenes, but also enable us to see similarities among objects, arising from the partial assemblies that they share. We have a strong tendency to see faces in many objects, if they have features that in any way resemble the organization of a face.

The increasing complexity of downstream areas also begins to explain the kinds of abstract relations we readily identify in objects and actions. We drop an object, see it fall and hit the floor, and we hear the sound of its contact. From the statistical regularity of these events we construct ideas of causality: the release causes the fall; the fall causes contact with the floor; contact causes the sound. We build a whole "folk physics" of similar abstractions about simple physical interactions. Similarly, we relate spatial locations: if you head north, and then turn right, and right again, you're heading south. In a related fashion we learn abstractions about social interactions: if someone smiles we infer they're pleased with something; if they grimace they must be hurt; if they cry they may be sad. The deeper we progress through cortical pathways, the more we arrive at hierarchical stages that synthesize many percepts into mental constructs that are recognizably cognitive. Proceed far enough, through the pathways of a sufficiently large brain, and we come upon regions carrying out uniquely human mental operations.

LIBRARIES AND LABYRINTHS

Picture a library with books arrayed in sections. Depending on the arrangement, and the sought-after topic, different paths through the library will be traversed. If we were to mark the paths of thousands or millions of visitors, we could then compile the statistics of traversal—which paths are walked most often from the entrance, but also where people tend to go between various sections (e.g., how many go from Travel to Fiction, or from Travel to Reference, or from Bestsellers to History).

This is one integrative look at the brain; memories are stored in various places according to categories or "specializations." Retrieval of those memories involves traversing to their locations and activating them. Storing new memories likely entails recognition and retrieval, during the process of "shelving" a new "book."

One more codicil: memories are not stored in their entirety, as books on a shelf. Rather, during traversal of a brain connectivity path, stages along the path add to the reconstruction of the memory: the memory becomes "assembled" incrementally along the path.

It is interesting to note that brain mechanisms are thus not much like any of the typically invoked analogies—they are not like telephone lines, not like the internet, not like computers. They're more like a scavenger hunt, in which a prescribed path is followed, clues are picked up and assembled, and later clues might be instructions to go back and pick up prior ones.

The great neuroscientist Sir Charles Sherrington, who we mentioned in chapter 6, described the brain as an "enchanted loom," where "millions of flashing shuttles weave a dissolving pattern." Indeed, if we had to have a technological analogy, brains might be something like combinations of cotton gins and old-fashioned spinning wheels—traversing fields, picking up individual raw material, assembling it and weaving it into threads of memory. Again, progress is not unidirectional. Some threads might induce movement backwards, down other alleys, to pick up further material from different locales to produce the final product. Similarly, when we see a flash of red and a gentle curve, we might think it either an apple or a sports car, and those realizations can trigger "backward" activation, looking back through the visual field to target further identifying information (a stem or leaf, a wheel or fender).

To combine the metaphors, imagine a library in which you walk through prescribed labyrinthine paths, picking up words and pages, assembling the book incrementally as you proceed.

The final "stage" at which the "book" is finally assembled, is not then where the book "resides"; rather, its parts are littered along the path to that final stage, the endpoint or turning point of an assembly path. So to arrive at the book, you traverse the library, reconstructing the book.

As we mentioned, in brain imaging studies, we see a particular area that is differentially active when, say, a face or a house or a forest is seen. We've come to call these "face" areas and "house" areas and "place" areas, which they in some important sense are, but we might even better think of them as either brain path "endpoints"—final stages at which we arrive when assembling a recognition memory of a face or a house—or as path "intersections," where two or more separable paths, containing the constituents of the memory, finally converge to enable the final re-construction or re-presentation of the face or house.

If that's so, then what fMRI and related imaging techniques might be measuring are those stages along brain paths—possibly endpoints and intersections—where crucial or final assembly takes place. The paths leading to those highlighted locales might also be active during these reconstructions, but might contain variation that renders them slightly less than statistically significant. Thus the (reliable) endpoints and intersections may show up prominently in imaging studies, leaving out the (more variable) paths that lead to them.

GRAMMARS OF THE BRAIN

The memory structures being built inside these systems have a recognizable organization. At each processing stage along a path, as we have seen, sequences of categories are constructed. These are nested hierarchically, such that a single category at one stage may itself be part of another entire sequence of categories.

These structures can be expressed in terms of grammars—just like those learned in school. This is a computational formalism, a code, that captures the crucial characteristics of these hierarchical brain representations. Here is a simple instance:

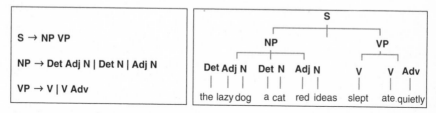

Figure 8.8   Successive organizations of neurons such as those illustrated in figure 8.7, create internal responses not just to images, like faces and houses, but to arbitrary signals. Progressing far enough downstream, the same brain mechanisms active in perception can organize complex sequences of features into linguistic structures.

This is an instance of a linguistic grammar; one used to describe sentences in English. As we'll see, brain grammars follow the same rules as these linguistic grammars, but brain grammars can be used to describe more than sentences; they can represent sights, sounds, and concepts.

The grammar in figure 8.8 says this: that a sentence (S) is composed of a sequence consisting of a noun phrase (NP) followed by a verb phrase (VP). In turn, a noun phrase is a sequence that consists of an optional determiner (a, an, the), an adjective (also optional), and a noun. A verb phrase is a sequence composed of a verb followed by an optional adverb. A standard way to write these grammars is on the left. On the right is an illustration of the same grammar drawn to show it clearly as sequences of categories.

Note that this particular structure generates only very simple sentences: A lazy dog snored contentedly. My computer crashed. The game started promptly. Her watch stopped. The bird soared majestically. His gun fired loudly.

The sentences are simple, but the most important thing about them is this: *we'll never run out.* You've got thousands of English words that can fit in each spot in the sentence; combining those will generate far, far more sentences than there are minutes in a century. This single, simple grammar, which takes just three short lines, generates more sentences than we could use up in a lifetime.

We have hypothesized that our brain circuits use the same underlying mechanism for vision, and for action, and for thought, as they do for language. As we've seen, the brain constructs these hierarchical sequences of categories throughout. Starting from purely perceptual information such as visual and auditory data, they build up hierarchically to representations in the brain of complex entities such as faces, places, houses, cats and dogs. And proceeding further, as more and more association areas are added, they continue to build ever more abstract relations among memories; relations such as movement, or containment, or ownership. Proceeding far enough downstream we arrive at almost arbitrary and abstract combinations of thoughts. Starting with just these internal entities, hierarchical sequences of categories can represent the entire panoply of our experience with the world.

CHAPTER 9

FROM BRAIN DIFFERENCES TO INDIVIDUAL DIFFERENCES

We have noted how similar all mammal brains are, and how those brains give rise to shared mechanisms—mammals all think pretty much alike. Yet we also find individual differences; even brothers and sisters have their own unique thoughts and manners. Perhaps we all start with similar brains, and we are molded into differences by our separate experiences in the world. But our brains themselves really do have internal differences. Genes build brains, and each of us has slightly different genes. Might we then be born with innate differences? This is a flare topic. It has been used—and misused— to intimate that different groups may have different intrinsic abilities; that black humans may be somehow intrinsically inferior to white humans, who may be somehow intrinsically inferior to Asian humans. What is the basis for these inflammatory claims of races and racial differences? How can we sort through them with what we know about brains?

Let's start with some individuals with notable differences. Kim and Les are middle-aged men, and Willa is an 18-year-old woman.

Les

Les was born prematurely, and with complications. He was given up for adoption at birth. He was blind, and had apparent brain damage, and was extremely ill. He was expected to live just a few months. A nurse-governess at the hospital in Milwaukee named May Lemke, who had already raised her own five children, took Les home, assuming that she would provide comfort for his short life.

But under his adoptive mother's care, he lived on. He developed an impressive memory, and would often repeat long conversations, word for word, including the different intonations of different speakers.

One night in his teens, he apparently heard the theme from Tchaikovsky's Piano Concerto No. 1 on a television program in the house. Later that night, his adoptive parents were awakened by the sound of that concerto. Initially thinking that they had left the TV on, they instead found Les, at the piano, playing the piece in its entirety, from memory.

He had never had a piano lesson.

Les Lemke now plays regular public concerts in the United States and abroad. He still has never had a piano lesson.

Willa

Willa's genome has a set of deleted sequences from about twenty genes, all on a single chromosome (7). Her rare condition, which occurs once in about 10,000 births, is called Williams Syndrome. Her brain is about 15 percent smaller than an average brain. At age 18, Willa functions at roughly the level of a first grader, barely able to perform normal adult activities. She can't drive a car or use a stove, and she requires supervision for the simplest tasks. But she can interact; she can use language more expressively than many of us. Here is an example of a spontaneous comment from Willa, describing herself: "You are looking at a professional bookwriter. My books will be filled with drama, action and excitement. And everyone will want to read them. . . . I am going to write books, page after page, stack after stack. I'm going to start on Monday" (Bellugi et al 1994). Willa routinely shows extreme linguistic fluency of this kind, and is able to produce richly imagined fictional stories, and to compose lyrics to songs.

Kim

Kim was born with a number of unusual brain features, including a lack of two fiber bundles (the corpus callosum and the anterior commissure) that are typically very large in most humans, and usually serve to connect large regions of the right and left sides of the brain.

Kim can read a full page of text in about 10 seconds, or an entire book in an hour. He can remember all of it, and currently can recall upon request any part of thousands of books from memory, including several telephone books. He has a number of other skills as well, from music to arithmetic calculation.

By some measures, he is a superman, capable of mental feats that most humans struggle with. By other measures, he is impaired: he has difficulty with abstract concepts, with social relations, with events that most of us think of as "everyday life." (His meeting with the writer Barry Morrow provided inspiration for the character Ray Babbitt in the movie *Rain Man*.)

The genetic differences between you and these individuals is small; accidental changes to a few genes. Each of these three people has challenges in dealing with the everyday world. Yet each of them has some ability that most "typical" people never possess—unusual musical ability, remarkable verbal imagination, superhuman memory.

If a slight change to a few genes can make manifest these powers in individuals, then it is likely that these same powers are hidden, dormant, in us. Their brains simply do not differ from ours by much. The fact that these small differences can unveil remarkable capabilities suggests that these capabilities are not actually very different from our own.

If you're like the average reader, you can't read this (or any) book at the rate of one page every 10 seconds. You can't memorize thousands of books. You can't play a concerto after one hearing, nor after a hundred hearings. These folks, with their slightly different genes and brains, can.

The existence of these individuals is an irrefutable demonstration that there are some relatively simple rearrangements of "typical" human brain structure that can give rise to abilities—often valuable, highly marketable abilities—that the rest of us simply don't have.

In their cases there appear to be trade-offs of some abilities for others. Possibly the unusual capabilities (high verbal and social abilities for Willa, musical ability in Les, memorization in Kim) are achieved specifically at the expense of other faculties. We'll return to this observation in a bit—but first, let's explore the implications of these rare abilities.

Apparently, brains can be slightly rewired to make memorization effortless. Apparently, slight modifications of the organization and pathways in a brain may enable amazing musical abilities. Apparently, verbal skills can be flipped on like a switch. If these abilities arise, without prompting, from individuals with slightly different brains, it may be strongly argued that the abilities are intrinsically in the design patterns for our brains, already there, ready to be unveiled, with just a bit of modification.

What kinds of changes might be involved?

BRAIN PATHS

Brain areas are wired to communicate with each other via cable bundles: tracts consisting of many axons that traverse large swaths of brain tissue, connecting brain areas to each other—some neighboring, some quite distant.

Some of these brain paths are readily identifiable when a brain is dissected. Axon tracts are "white matter," named for their greater reflectivity than that of cell bodies. Some are so prominent that they were named early on by neuroscience pioneers: the corpus callosum, arcuate fasciculus, corticospinal tract. More have been discovered by neuroanatomists via careful staining and tracing of fibers. Recent advances in brain imaging are enabling us to create virtual traces of axon tracts in the living human brain, via a plethora of "tractography" techniques. We are coming to uncover the actual pathways through the brain.

It may be thought that, with all of our technology, all the paths of the brain are already mapped, named, and understood, but this is far from true. In fact, anatomy—the study of the actual structure and organization of the brain—has fewer adherents, and less funding, than many other topics in brain science, from genetics to cognition; yet all other fields of research on the brain are dependent on anatomy.

As we saw in the previous chapters, successive brain areas produce increasingly complex combinations of their inputs. Brain signals are sorted and conveyed through appropriate routes, becoming incrementally elaborated at each stage of processing. Brain stations early in the process specialize in simple initial features; signals eventually arrive at far-downstream stages specializing in assemblies of those features into regularly occurring patterns. These patterns are our memories, shaped by slight changes to synaptic connections between brain cells, in ways corresponding to experiences we have had. "Specialist" regions arise from these experiences; if you have seen many instances of pine trees, you will process new pine tree images differently than you would have before those prior experiences. From images of faces and houses, to sounds of speech and music, these patterns are sorted by specialists deep in our brains.

The various pathways through the brain define the functional assembly lines through which percepts and memories proceed. The axon tracts through the brain determine the operational paths through which signals will be shuttled, and determine the successive stations at which different brain areas will contribute to the assembly of a memory.

These pathways can be traced in living humans by very recently developed "tractography" methods. In particular, a set of methods

referred to by the somewhat abstruse term Diffusion Tensor
Imaging, or DTI, enables scientists to trace the paths of axon tracts
throughout a living brain, thereby entirely reconstructing the
anatomic connectivity of brain areas to each other. Some are
illustrated here.

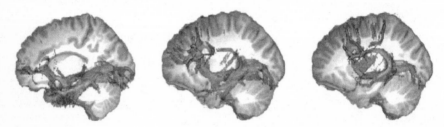

Figure 9.1   Structural pathways through a human brain, educed
via diffusion tensor imaging. (from Anwander et al. (2007); used by
permission)

These images show the connection pathways that exist among
different areas of the brain. In functional imaging experiments, we
are able to trace the areas that are selectively activated in the brain
during particular tasks such as reading, recognizing certain objects
or places, and observing emotions in faces. By this method, we can
combine information about paths and functional activation. If we
can trace the pathways through which activation travels from one
brain area to another, and we can see what stations along those
paths are selectively activated in particular tasks, we can begin to
glimpse entire assembly lines at work.

Such studies could, in principle, enable us to ask a loaded
question: how might individual brains be differentially wired for
different predilections?

And so, these answers might then follow:

- Different groups of people have different mixtures of genetic
  features.
- Slight gene changes can give rise to differences in brain path
  connectivity.

- Differences in brain paths can affect the ease with which certain behavioral functions may be performed.

The implication is clear: innate brain connectivity differences can lead to individual and group differences, with disparate talents arising from various connectivity patterns.

We perhaps should note that genomic control of connectivity in the brain is likely to be highly indirect. Several years ago we removed a pathway in the forebrain of an immature rat, and watched what happened in zones in which that path normally terminated. Amazingly, other connections to this zone accelerated their growth and in just a couple of days took over all the territory normally assigned to the now-missing input. We had, without meaning to, rewired the brain. These experiments, and many more that have replicated and extended them, demonstrate that growing pathways do not receive direct specific genetic instructions about their ultimate size and destinations. Pathway growth may be more like a gold rush than a precisely-orchestrated engineering job. Genes affect this process by loosely specifying how many neurons will arrive in a particular area of the cortex, and specifying when they will arrive during development; thus mutations affect connectivity only indirectly.

BRAIN TRACTS AND DIFFERENTIAL ABILITIES

Recent studies have examined some specific predictions of this hypothesis. A number of laboratories have used DTI to identify the differential connections of people with measurably different abilities. One example that has been repeatedly studied is reading. From people with specific reading difficulties, such as dyslexics, to unusually fast and accurate readers, there is a broad and apparently near-continuous range of reading abilities. Scientists have measured people's brain connection pathways and compared these with their reading abilities. What they have found is exciting, and potentially troubling.

It turns out that reading ability can be correlated with details of the connection pathways linking particular brain areas. An area toward the back and left of the brain is differentially active when people recognize the visual shape of words; another area, toward the left front of the brain, is active preferentially when people recognize the sounds of words, such as rhymes. These two areas, together with a primary axon tract that connects them, the superior longitudinal fasciculus (SLF), are more weakly connected in dyslexics than in non-dyslexic readers. There is actually a whole constellation of slight differences that have been reported between dyslexic and non-dyslexic readers; this discussion illustrates just one prominent component.

Moreover, tests have been run not just in the separate groups of dyslexic and non-dyslexic readers, but also in ranges of readers exhibiting reading skills from very good to poor, including intermediate-level readers. They found that the diminution of these brain paths was correlated with reading level: the best readers had the strongest connections between these brain areas, and these areas were least connected in the weakest readers. So the relation between these brain paths and reading was not just one of intact readers versus those with a specific deficit—rather, the relation is a continuous one: the more connected these brain areas, the better the reading abilities.

In any finding of this kind, the correlation (weaker readers have weaker connections between two reading-involved brain areas) cannot be immediately imputed to causation: that is, do weaker connections cause impaired reading, or do poor reading abilities cause these connections to weaken—perhaps due to less reading practice? In general, questions of this type require careful experimentation to address.

Scientists therefore studied these correlations in children aged 7–13, who had far less experience with reading than adults. The correlation still was shown to hold, suggesting that differential practice or reading experience was less likely to cause the connection changes—rather, the connection changes were likely the cause of the differential reading abilities in children and in adults.

NATURE AND NURTURE

This could lead to downright disturbing inferences. Is it possible that we are born with genetic predispositions that affect the strength of connection tracts in our brains, and that these in turn predetermine—predestine—our abilities for the rest of our lives?

The truth is quite different. Genetic predispositions are just that—tendencies that influence brain growth, not absolutes that dictate it. Indeed, it has routinely been found that the genetic features we are born with are likely to be responsible for about half of the differences between one individual and another—with the other half arising from non-genetic influences, which include environment, parenting, siblings, peers, school, and nutrition, to name but a few. Comparative studies have been carried out of twins separated at birth, non-twin siblings (biological brothers and sisters), and adoptive siblings. Separated twins share all their genetic features but none of their environmental influences; biological siblings share some of their genes and much of their environmental influences; and adoptive siblings share their environment but no genetic material. Statistically, about half of the similarities and differences among these groups can be accounted for by their genetic backgrounds, and the remaining half cannot, and must be attributed to environment. Genetic predisposition is a tendency, but it clearly is not predestination. It is likely that brain pathways are influenced in equal measures by nature and by nurture.

Again, the effects may be quite indirect. Studies of identical twins are often interpreted strictly in terms of genes and brains, but of course twins share body types, hormone levels, visual acuity, and countless other variables, all of which affect the way the world treats them. How a child gets along in school is influenced by their height, weight, athleticism, skin color; and how the child gets along will certainly influence his or her mental makeup. This is one reason that some scientists find claims of inheritance of cognitive skills and talents to be only weakly supported.

Moreover, brain pathways may underlie the entire diverse spectrum of individual abilities. These pathways, influenced by genes and environment, play a part in specifying differential abilities in music, in athletics, in affability—in a broad range of characteristics that make us who we are. Far from determining a linear ordering of individuals who will "win" or "lose," differential brain path arrangements can grant a range of talents and gifts, leading in diverse directions, helping to generate populations of individuals each with unique traits to add to the human mix.

CHAPTER 10

WHAT'S IN A SPECIES?

We humans are loners: we are the only surviving species of our own evolutionary group.

It's highly unusual. Animals of most every other species have living "cousins," closely related species, derived from common ancestors.

For instance, lions are referred to as "*Panthera leo*," referring to their "genus" or generic category *Panthera* and their "species" or specific category "*Leo.*" There are three other living relatives in the genus *Panthera:* tigers (*Panthera tigris*), jaguars (*Panthera onca*), and leopards (*Panthera pardus*). There are also many recognized subtypes or subspecies, such as *Panthera leo hollisteri*, or Congo lion, and *Panthera leo goojratensis*, Indian lion, which can inter-breed but whose geographic separation makes that impracticable, resulting in diverging characteristics.

Humans are the apparent stepchild, uniquely excluded from this family arrangement. We can be referred to as "*Homo sapiens sapiens*" denoting our generic category *Homo* (Latin for man), our species *sapiens* (Latin for wise, intelligent, knowing), and our subspecies, further emphasizing our sapience, and denoting our possible divergence from other subspecies, now extinct.

Just as with lions and jaguars, there are a number of other species members of our genus: *Homo habilis* ("tool-using man"), possibly the earliest true member of the *Homo* genus; *Homo erectus* ("upright man"); *Homo neanderthalensis* (from the Neander valley in Germany, where the first Neanderthal bones were found).

One difference is clear: there are four living species (lions, tigers, jaguars, leopards), and multiple subspecies, all sharing the genus *Panthera*. All of our relatives—every other member of our genus, species, and subspecies—are extinct.

A simple chart identifies estimated dates when our departed relatives lived.

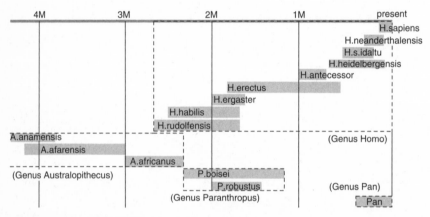

Figure 10.1   Approximate timeline of human relatives. Australopithecines evolved both into members of more ape-like creatures Paranthropus, and into increasingly human-like species of genus *Homo*. *Homo* eventually evolved humans and many now-extinct relations.

Apparently, throughout most of hominid history, rarely did more than one or two known human-like ancestral species coexist. After the hypothesized split between more gracile, human-like (*Homo*) and more robust, ape-like (*Paranthropus*) genera, there are few overlaps in time. *Homo erectus* deserves special note: this hardy soul apparently held down the fort, mostly alone, for more than half a million years—roughly a hundred times the duration of all recorded human history. We *Homo sapiens* come from a long line of loners.

The alternative possibility to the notion of *Homo* species isolated existence of course exists: that the few fossils that have been found may represent more species than are typically attributed to them—and that there may have been many additional species whose fossil remains have not yet been identified. Fossils are, indeed, fewer and rarer than one might think. For instance, the very first fossil of an ancient chimpanzee, believe it or not, was just found in 2005; before that, no ancient chimp fossils had been uncovered. If hypotheses of species were solely dependent on identifiable fossils, our view of family trees would be far different than it is. Fossils are all too easily damaged by time. There may have been other species out there—possibly many others—whose fossils are either irretrievably buried or destroyed.

DEFINITIONS

Disagreements abound regarding family trees of this kind. In one camp, it is argued that chimpanzees and possibly gorillas ought to be categorized in the genus *Homo*, rather than in the genus *Pan* or genus *Gorilla*, respectively. In another camp, there are proposed subspecies of *Homo sapiens*, including *Homo sapiens idaltu* ("elderly wise man") and *Homo sapiens neanderthalensis* (assigning Neanderthals to a subspecies of *Homo sapiens*, rather than a separate species in genus *Homo*). We then are the subspecies *Homo sapiens sapiens*. Do Neanderthals and Idaltus share our species, or just our genus? Are they our sisters or distant cousins? Scientists in Germany have recently decoded the Neanderthal genome, and this and other findings provide evidence both for and against the inclusion of Neanderthals in our species.

These uncertainties highlight one of the central difficulties of species designation: that of naming and correctly classifying groups of animals. Although we tend to think of these assigned names and categories as scientifically tested and validated, the reality is often somewhat starker. These categories of genus and species, all collectively referred to as "taxons" or "taxa," i.e., units of taxonomy, are sometimes applied according to the idiosyncratic proclivities of particular researchers, rather than according to testable hypotheses that can be independently validated or invalidated.

As is often the case, such concerns may not be easily changed from within, but some in the field have articulated the problem. Recently, the prominent paleoanthropologist Jeffrey Schwartz wrote tellingly of some scientists' "bias against recognizing taxic diversity in the human fossil record" (Schwartz 2006): "in contrast to the typical paleontological experience of discovering new taxa as new sites are opened or as already-known sites continue to be excavated, it is not uncommon to find paleoanthropologists arguing against the possibility that hominids could have been as speciose in the past as undoubtedly appears to have been the case for other groups of organisms."

In other words, scientists have no trouble identifying multiple species in most generic categories of organisms (such as panthers), and yet, despite continuing human-like fossil discoveries, at sites new and old, the number of species assigned to genus *Homo* continues to be small. Two possibilities: there are additional, unacknowledged species within the genus *Homo*, or there are indeed few species in our genus. Either is a fact that cries out for explanation.

A central difficulty is this: there are many specific features (relatively big brains, certain kinds of teeth, grasping opposable thumbs, and many more) shared among all members of genus *Homo*, and for that matter among the entire larger category, the family Hominidae, which includes chimps, apes, gorillas, and orangutans, as well as us. There are also features (really big brains, certain kinds of tool use, etc) that appear to occur only in genus *Homo* and not in any other genera of the Hominidae family, and even features (construction, language use) that occur only in the species *Homo sapiens* (and possibly only in the subspecies *Homo sapiens sapiens*).

The first question, then, is what features to use as dividing lines among species. If we "lump" lots of variants together, we may get very broad categories that contain many very different subgroups; if we "split" variants wherever they occur, we may get multiple categories whose members barely differ from each other.

FALLACIES OF THE NOTION OF RACE

One insidious track that sometimes creeps into discussions of these points is the question of superiority. Put bluntly, if supposedly "primitive-looking" fossils such as Neanderthals are separated out into different species, it might be possible for some to argue that perceived differences among living humans are biologically, and thus perhaps evolutionarily, significant: that some are better than others.

Social implications of such inferences are readily dispelled. First it's crucial to recognize the difference between separate and inter-breeding pools of individuals.

Separate pools. Members of "separate" gene pools (think different species) are those that do not interbreed. Differences between such groups (e.g., lions and jaguars) are thus generated independently (by genetic variation) and are selected semi-dependently (by competition for resources and niches).

Interbreeding pools. In contrast, members of overlapping gene pools (think variation among members of a single species) do interbreed. Differences among these individuals are different in kind from the differences between members of non-interbreeding groups. In particular, these individual differences are subject to selection pressure, since members of these groups might compete with each other not only for the same niches, but also for "procreation rights" with partners.

Scientists have learned that these are not strict categories. By roughly 5 million years ago, our ancestors (i.e., those whose genes we would eventually inherit) had established separate genetic characteristics from primates who would become the ancestors of chimps. Before that time, these groups had constituted a single interbreeding gene pool—the forebears of both humans and monkeys. However, after this separation between human and chimp ancestors had been established for almost a million years, there is evidence of a recurrence of interbreeding between these ancestral entities. When this finding was announced, in 2006, it came as a surprise to all; it had been widely assumed that separate populations would remain genetically incompatible.

If it were true that some genetic characteristics were "better," one might conceive of a draconian policy to "improve" humanity by eliminating individuals with "inferior" traits—either by killing them or by selectively denying them the right to procreate.

Of course, we know virtually nothing about which traits are "superior" or "inferior," let alone how such traits might interact with each other in interbreeding populations. As discussed in chapter 3, some traits are inextricably linked with others, since the genome compresses all of our features into just 20,000 or so genes, by mechanisms whose principles are still barely apprehended. In light, then, of our still spectacular ignorance about genes and populations, embarking on a plan to "clean it up" would be laughably—or tragically—misguided.

Humanity has a history of such attempts, in all their ignorant splendor. The world abounds with those who would systematically enslave or murder all those with certain physical traits that differ from their own: from Nazis to Rwandans, from European colonizers of native populations around the world, to small local groups who deny the right to exist to their slightly different neighbors (who, to outsiders, are typically indistinguishable from their oppressors).

It is surprising to most Americans to learn that, in the very recent past, there was a social movement called "Eugenics," sometimes defined as the "self-direction of human evolution," which was widely embraced by scientists early in the last century, including such otherwise-notable personages as Alexander Graham Bell, George Bernard Shaw, and Winston Churchill, among many others. It was largely a program to "encourage" certain "desirable" traits and to "discourage" traits labeled "undesirable."

The methods of eugenics stopped short of murder. But they included, amazingly, mandatory sterilization for people who had traits the state declared undesirable. The lists of such traits included various mental illnesses (ignorantly typed as incurable), certain diseases such as tuberculosis (ignorantly argued to be heritable), and even "chronic pauperism," i.e., people without ready cash.

It's sobering to look at eugenics in any detail. Individual states in the United States, beginning with Indiana in 1907, created laws mandating compulsory sterilization for those labeled undesirable. Indiana's example was followed by dozens of other states and countries around the world. Who was to be sterilized? The law called out "confirmed criminals, including the categories of 'imbeciles' and 'idiots'." Indiana's law remained in force until 1921, when it was declared unconstitutional—and then reinstated by the legislature in 1927. The law was eventually repealed once and for all . . . in 1974.

RACES VERSUS GENE POOLS

Figure 10.2 is a schematic depiction of the genetic makeup of a select group of hypothetical individuals, together with their phenotypic characteristics, i.e., aspects of their visible appearance.

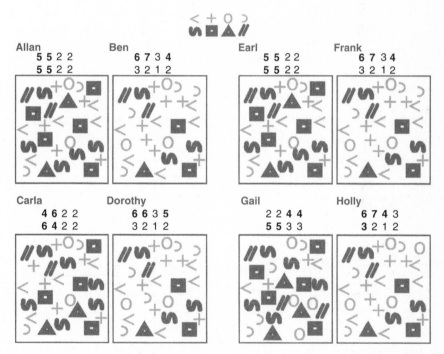

Figure 10.2 Schematic diagram of genes, and the features of the organisms they construct. Eight individuals are shown, each with different "gene" arrangements. Numbers indicate the number of each genetic component (+, 0, <, etc.) contained in a given individual's genome. Boxes show some anatomical features of the resulting individual. The example illustrates that particular visible features are not good indicators of underlying genetic characteristics: individuals with similar features may nonetheless have different genetic makeup, and, reciprocally, individuals with different features can have very similar genetic makeup.

The example shows two different ways that genetic variation among humans can occur. At the top are eight hypothetical genetic features, each denoted by a particular symbol: $<, +, o,), \sim, \square, \Delta, //$. Some of these genetic traits show up phenotypically, that is, in the body, such as height, genetic obesity, or color of hair, eyes, or skin. But many others do not manifest themselves in overt features. For instance, some genes control certain aspects of body chemistry, some to do with our eyesight, some to do with predisposition for certain diseases. The question is whether these genes all correlate with each other. That is, if someone has a certain skin color, how much does that tell you about his or her other genetic features? The answer is, it tells you almost nothing.

For our example, we assume that individuals carrying any of the physically darker features ($\sim, \square, \Delta, //$) may tend to exhibit differentially darker skin color than those carrying fewer of these features.

In the rest of the figure, each box is an individual person, carrying some particular mix of these genetic features. For each individual, the number of copies of each of the eight genetic features is shown in the list of numbers, together with the resulting "appearance" of an individual with that genetic profile, in the box. For instance, at the top left, Allan has 5 copies of the "$<$" gene, 5 of the "$+$" gene, two each of "0" and ")", five each of "$\sim$" and "$\square$", and two each of "Δ" and "//").

On the left are four individuals, two of whom (Allan and Carla) have a "dark" appearance and two of whom (Ben and Dorothy) have a "light" appearance. In these four people, the underlying mix of genes is correlated with dark versus light appearance: the two left-hand individuals, Allan and Carla, both possessing dark features, have similar patterns of genes (5522, 5522 for Allan, and the similar 4622, 6422 for Carla), and the two right-hand individuals, Ben and Dorothy, both light-skinned, have their own similar genetic patterns (6734 3212, and 6635 3212, respectively). Thus these individuals might be drawn from populations in which there actually exist "races," i.e., underlying genetic profiles that are recognizable by visible traits.

On the right is another set of four individuals. Again, two appear "dark" (Earl and Gail) and two "light" (Frank and Holly). But here, the two left-hand individuals are both dark-skinned but have very different underlying genetic patterns: Earl's genes are 5522 5522, whereas Gail's are the very-different 2244 5533. The two right-hand individuals, both light-skinned, also have very different underlying genetic patterns from each other, again despite both having the side-effect of a light overall appearance. This is a population, then, in which visible traits such as skin color do not correlate with gene pools. That is, if one attempted to guess the overall genetic makeup of an individual from his or her visible traits, the results would be very unpredictable.

The population on the left is a cartoon; that on the right more accurately reflects the reality of human traits. We have highly overlapping gene patterns. A few of our genes yield light or dark appearances, but these few genes are not correlated with the rest of our genes.

Pick a set of people who all share a given trait: those with the same skin shade, or the same eye shape, or nose shape, or ear shape, or the same height range, or hair or eye color. Within that single group, or so-called race, there will be just as much genetic variation among all the rest of their genes, as there is in the entire human population. The converse is also true: sub-groups of the population that have particular genotypes, i.e., whose genetic patterns closely match each other—members of an extended family, for instance— nonetheless exhibit a wide variety of overt traits, which is why brothers, sisters, and cousins often look, on the surface, very different from each other despite being genetically very related.

CHAPTER 11

THE ORIGINS OF BIG BRAINS

We have discussed what it means to speak of humans as a species, and a range of contrasting ideas about the types of brain changes that resulted in the first sparks of our intelligence. Now we turn to the continued growth of the brain, leading to human-level intelligence. We begin with some seldom-asked questions: Why did human intelligence stop with us? Could it go farther? Could our brains add material, and add function?

Addressing the questions starts with examination of our closest relatives, all of whom are extinct. The story begins in Amsterdam, more than 100 years ago.

After Darwin published his work, the idea of evolution took off like a wildfire in the world of ninteenth-century European intellectuals; and nothing fanned the flames more that the discovery, in 1887, of the Neanderthals. Here unmistakably was something from a long ago time that was almost, but not quite, human. The missing link in the progression from the imagined ape-like ancestor to modern man had, apparently, been found. But one of the great biologists of the time, Rudolf Virchow, nearly dealt a death blow to speculation about the Neanderthal. He pronounced that the skeleton was simply that of a modern man who suffered from a terrible case of rickets. Virchow's normally keen scientific intuition had been blunted by his visceral hatred of the idea of evolution, a pathological condition that arose from his remarkable insight that the theory would eventually be used to promote extreme right-wing agendas, some of which we have discussed and dismissed in the previous chapters. His heart was in the right place, but his conclusions were wildly wrong. Virchow's preeminence was at the time being challenged by the younger and far more charismatic Ernst Haeckel, the author of the interesting but still very wrong idea that the evolutionary history of a species was echoed in the development of an individual. This theory was to distort the biological sciences right up to the present day.

Haeckel embraced the Neanderthal remains and through a whirl-wind tour of public lectures and publications used them to describe the "true" missing link; he named this creature 'Pithecanthropus', the ape-man. This half-way human, Haeckel explained, walked upright and in other regards looked human but lacked the brain power for speech, and thus could not escape the bestial aspect of its evolutionary inheritance. Armed with this graphic image, all that was required to secure the entire theory of evolution, to explain the origins of humanity itself, was someone with the means and determination to find the bones of Pithecanthropus. These words eventually found their way to just the person needed for Haeckel's wild, insanely romantic quest.

Eugene DuBois, a successful member of the small but influential Dutch medical and scientific establishment, saw in Haeckel's vision the chance for immortality. Being a thoughtful man, he laid out a set of arguments about Pithecanthropus' final resting place and, happy answer, concluded it must be located on one or two of the islands that constitute the Dutch East Indies (modern Indonesia). All that remained was to join the army, pack up the wife and family, get assigned to the right islands, and start digging. Incredibly, this came to pass—DuBois unearthed the true missing link or, more accu-rately, one of the links leading from ape to human. Needless to say, his claim was the opening shot in an academic war that lasted the rest of his life, with some saying it was a giant gibbon, others that it was a human, and others siding with the discoverer's view that it was neither ape nor man.

Quite properly, the most bitterly contested battles concerned the size of the brain (as deduced from the cranium), both in absolute terms and in relationship to the size of the body. Pithecanthropus' brain indeed turned out to be above the range of the apes and safely below that of *Homo sapiens*.

From the ensuing research came a set of sharply defined mathematical descriptions of the true relationship between brain and body size in primates and other mammals. And from those descriptions, we gain a measure of the human brain, which, as we'll see, seems to violate these brain-to-body relationships.

BRAIN SIZE IN THE PRIMATES

We have shown the measure of the human brain: it is approximately 1350 cc, which is more than three times larger than that of our nearest living relative, the chimps. But chimps are smaller than humans, and perhaps their brains in turn are smaller than normal, or larger than normal. How, in more general terms, are brains related to bodies?

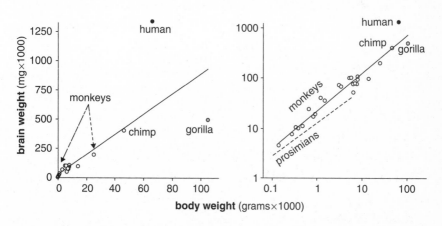

Figure 11.1   Relationships between body size and brain size in monkeys, apes, and humans. Left and right panels show the same data, plotted first on linear scales (left) and then on logarithmic scales (right) to spread the data more evenly. For most primates, brain size follows predictably from body size. Humans are a prominent exception to the rule.

A graph of brain weight vs. body weight for a set of twenty-seven higher primate species shows that *Homo sapiens* sits, as expected, way above the curve, but surprisingly that the apes, though near us in many ways, do not (figure 11.1). (This first graph has the unfortunate side effect of shoving many of its points into the lower-left quarter, with most of the space dedicated to just humans, gorillas, and chimps, since they are so much larger than most monkeys. A simple change enables these points to appear more uniformly, as in the right-hand graph. In this plot, each tick mark is actually ten times greater than the next lower mark. Instead of reading out 0, 10, 20, 30, . . ., we read out 0, 1, 10, 100, and carefully plot the size data in the appropriate locations between these expanding marks.)

The graphs make it clear that there is a tight relationship between body size and brain size, and also make the point that humans deviate from the expected brain size more than any other primate; our brains are about 2.3 times larger than would be expected from other primates. (We also see that the gorilla has less brain than expected for its size, a surprise that we'll return to later in the chapter.)

The dotted line on figure 11.1 adds information: the brain and body relationship for lower primates, the "prosimians" such as the bushbaby. Their brain to body ratio is more like that of all other mammals, while the monkeys and apes have somewhat larger brains for their bodies. What this means for humans is that not only is our brain about 2.3 times larger than would be expected for apes and monkeys, but it's even larger, about three times too large for its expected size when compared to all other mammals.

Why are brain and body size so closely linked? One simple explanation might be that the sensory and motor demands of running a bigger body require extra brain space. For example, apes have a much larger skin surface area with a correspondingly larger number of touch receptors than do monkeys; all of these receptors have to be represented in the brain, something that can only be handled by expansion. This is correct, but it is only part of the answer. When we look inside the brain, we'll find that sensory and motor areas constitute only a fraction of the expanded brain.

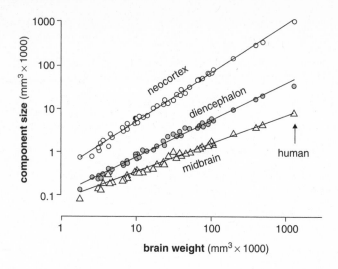

Figure 11.2   Regions and structures within the brain are highly predicted by the overall size of the brain. Humans (far right) are no exception: the size of our neocortex, diencephalon, and midbrain are all what would be expected from any primate that had our brain size. Note that the slopes of the lines are significantly different from each other. This means that the relative proportions of brain regions change as the brain grows larger; the cortex is 10 times larger than the mesencephalon in the small-brained bushbaby but 120 times larger in the human.

Figure 11.2 shows the relative sizes of internal regions of a brain, compared to the overall size of that brain. The three brain divisions plotted are the relatively ancient midbrain, the diencephalon, and the neocortex. What we see is that lower areas, primarily involved with sensory and motor processing, do increase; but neocortex increases much more, per brain size increase, than these lower areas do. As the brain gets bigger, then, most of the increase is not dedicated to sensory and motor needs, but to new neocortical areas. Big brains add plenty of material beyond what apparently is called for just based on differences in body size.

Another key fact shown by this graph is that all these brain areas scale very predictably. Knowing just the size of a primate's brain, from a bush baby to a human, we can determine the size of each of its internal divisions with great accuracy. If brain areas were changing due to external evolutionary "pressures," it would be a great coincidence for all those changes to fall right onto these lines. Instead, the evidence strongly suggests that brain parts are changing in lock-step, indifferent to external circumstances as brains grow large.

Another thing to note is that human brain components, on the far right, are just as predictable in size and in internal component composition as those of other primates; our brains are assembled just as an unusually large ape brain is.

The sizes of individual brain structures (striatum, hippocampus, hypothalamus) are also highly predictable based solely on the overall size of the brain. Again, humans are no exception to the rule; given our brain size, the internal structures are the expected relative sizes. For many animals, some variability can be seen, possibly indicating minor effects of evolutionary pressure.

Within this overall pattern of concerted growth, some variations occur. Possibly a gibbon with a slightly larger-than-expected hippocampus had extra survival value, and those changes might have resulted in more gibbons with a large hippocampus, whereas this relation may not obtain for, say, a colobus monkey. Again, these are not Lamarckian ideas: evolution can't know in advance whether a larger or smaller hippocampus is handy. Rather, accidental variation occurs, and if the difference causes better average survival rates for the species in question, there will be a subsequent increase in the probability of that previously-accidental difference in future progeny. We're used to seeing such features in animals' bodies; the tree-dwelling gibbon has longer arms and legs than the terrestrial human. In general, accidental features that confer advantages have a better chance of surviving and having their salutary genetic characteristics passed on.

Even with individual species-specific variations, the brain stays remarkably stable, never straying far from the component sizes expected from the overall brain size. All primates use the same brain pattern.

In particular, the biggest brains have much more cortex than smaller brains. Moreover, a bigger cortex has within it much more association cortex than sensory regions. This gives different relative intelligence for different brain sizes.

Apes don't have brains any larger than expected for primates of their body size; an ape body is larger than a monkey body, and ape brains are correspondingly larger than monkey brains, as expected. But apes really are smarter than monkeys, in tool use, communication, planning. Scientific studies show that chimps, unlike monkeys, can acquire rudiments of speech, and they organize themselves into raiding groups to invade neighboring territory—the beginnings of coordinated war. All of this behavior, reminiscent of humans, is being generated by an expanded monkey brain, whose differences are just those expected of chimp-sized primates. Intelligence came to the apes not due to any particular selected competitive advantage, but as an accompaniment to their unusually large (for primates) bodies.

The same is true for us. Much of the human condition arises from the predictable construction of an extremely large monkey brain.

In our case, however, there is an extra jump, as we saw in figure 11.1. Not only do we have the large brain expected of a large mammal, but we have a brain three times larger still than that! How did we transcend the rules that determine brain size from body size?

BRAIN SIZE IN THE FAMILY OF MAN

About 6 million years ago an ape-like line of primates split with one of its branches leaving the tropical forests for the open savannah, a lifestyle change that favored upright posture and bipedalism. Recent evidence from molecular biology suggests that the breakup must have been hard, because, as we mentioned in chapter 10, the two groups went on interbreeding for hundreds of thousands of years. But eventually, say by 5 million years ago, there were two distinct animals, with the open-space explorers leading toward proto-humans and those staying behind resulting in chimps. It turns out that leaving was the better idea because the great forests that had covered Africa for tens of millions of years were shrinking, and taking with them the ancestral homeland of the apes.

There are a range of names used for the proto-human primates, those who split off from the chimps. They, with all their descendents, are often referred to collectively as the "hominids," a taxonomic family to which, as described in chapter 10, there is but one species left. Some hominids are our direct ancestors. Others may have branched off during the last few million years, leading to branches containing other relatives of ours, not direct ancestors. All of these other branches have died out.

As we mentioned in chapter 10, the fossil record of the hominids is uncomfortably small and uneven; there are precious few fossils that have survived long enough to be found by paleontologists. Going back about four million years we find *Australopithecus afarensis*, a very early hominid that stood perhaps 5 feet tall and weighed around 100 pounds. Yes, stood: they had human-like pelvis and leg bones to stand on, though they still had ape-like faces. But the brain, at roughly 440 cc, hadn't gone past the chimp level.

Afarensis' appearance, an ape-like head with a body beginning to take on human form, answers one of the most debated questions of nineteenth-century biology: Did bodies become human-like before brains did? Darwin believed that the answer was no: that the human mind evolved first, and the body followed it. He posed this as his hypothesis for human descent from the apes, and he was wrong. The "brain-first" hypothesis came from the belief that evolution was driven by selection pressures for the thing that most defines us: our intelligence. This was, perhaps, another instance of the irresistible fallacy: the notion that what is important to us must also be important to evolution. But current chimps and apes, and, for that matter, current rats, are every bit as evolved as we are: all have been selected as survivors. We happen to have outsized brains; they have other features that enabled them to survive. Evolution doesn't know, and doesn't care. It selects accidentally, and preserves those features that happen to enable survival. In our case, it selected first for upright posture, and only much later for large brains.

Brain size in the subsequent Australopithecus species, *Australopithecus africanus*, barely inched above that of their precursors, to about 450 cc, although their skulls exhibited a few more human-like characteristics, especially in bones around the jaw, which decreased from the larger ape structures to somewhat smaller size.

It took until about 2 million years ago for things to start changing again. That's when the genus *Australopithecus* apparently split off two new radiations, one more ape-like and the other more human-like. The former, referred to as *Paranthropus*, had skulls with thicker bones and large jaws; altogether more ape-like than *Australopithecus*. The second group, contemporary with *Paranthropus*, is the first creature recognized by paleontologists as an ancestor to modern man. This was the first member of what was to become the generic category *Homo*, to which humans belong. This first such species is called *Homo habilis*. They were still just five feet tall, and still had a quite ape-like face, but now had a brain that exceeded 600 cc. And that brain apparently crossed some threshold for intelligence, because *habilis* is the first species we know of that designed and used tools. In fact, he is named for this ability: *Homo habilis* means "handy man."

Tool use is a hotly debated topic. Present-day chimps with their 400 cc brains strip leaves from thin tree branches, and use these to probe insect dens; they also pick up rocks and use them to crack open nuts. By comparison, *Homo habilis*, with his 600 cc brain, would collect particular types of rocks, and would split them to create sharp edges, fashioning a knife that he kept, and used repeatedly, for slicing meat from bones. (Meat was just an occasional treat in a largely vegetarian diet.) Around 1.5 to 1.8 million years ago, one of the many offspring of *Homo habilis* showed extensive new characteristics. These hominids, which we now refer to as *Homo erectus*, had longer legs, making them between 5 and 6 feet tall; their bodies were a lot like ours, and their brains had leaped ahead again, to 800 cc, now almost twice the size of a comparable sized chimpanzee. They apparently discovered how to set fires, enabling them to stay warm, and to light up dark caves, and to cook animal flesh. Their upright carriage allowed them to walk long distances; their fires enabled them to brave cold places. During the next million years *Homo erectus* spread out of Africa and across Europe and to easternmost Asia (where one lay waiting to be discovered, millions of years later, by Eugene DuBois). During this period of *Homo erectus'* rapid expansion, they evolved extensively as well. By around 500,000 years ago, they had brains of 1,000 cc, which reach the low end of today's humans. They seemed to be on the road to great things.

Around a million years ago, recognizably different fossils begin to appear: *Homo antecessor* and then *Homo heidelbergensis*, each with somewhat increased brain size. *Homo erectus* stuck around through this period. Then, within the last few hundred thousand years, came *Homo sapiens*, apparently in various forms: *Homo sapiens idaltu, Homo sapiens neanderthalensis*, and today's humans, *Homo sapiens sapiens*. *Homo erectus* eventually disappeared, though the reasons are subject to debate: either they were supplanted by competition from the new larger-brained *Homo* species, or as is sometimes argued, they drifted into many different lands and each of these separate groups may have independently evolved into modern humans. If the first case is true, then we were all geographically close for much of our history, and likely diverged only in the past few tens of thousands of years; in the latter case, there may have been geographically separate peoples even longer ago. Either way, the evidence suggests that *Homo erectus* and early *Homo sapiens* overlapped in time. There are signs that the last members of *Homo erectus* were still walking around in Indonesia possibly as recently as 30,000 years ago. One wonders how encounters with the earlier species might have provided humans with fodder for myths about the "old ones."

The transitional fossils labeled as *Homo heidelbergensis* were not much different from *Homo erectus*. They were first discovered as just a jawbone, in sand pits southeast of Heidelberg, Germany, giving them their name, but subsequently have been found in Africa as well. These beings weighed a bit more, perhaps 125 pounds, and had somewhat bigger brains, in the low end of the human range, around 1,200 cc. They looked like classic movie "cave men."

Physical anthropologists generally agree that *heidelbergensis* evolved into the Neanderthals, surely the most abused creature since Wallace and Darwin launched the theory of evolution. Neanderthals appear in the record around 200,000 years ago and may have arisen earlier than that. They persisted until at least 30,000 years ago across a broad swath of Europe and the Middle East. Yet again, the big question is what happened to them. Were they out-competed by the most recent wave of fully modern humans, or did they become absorbed into them by cross-breeding? Inextricably linked to this question is the one raised earlier: whether Neanderthals should be included as part of our species, *Homo sapiens neanderthalensis*, or as the distinct, closely related species *Homo neanderthalensis*. Neanderthals were relatively short and apparently powerful. It has, at various times, been claimed that they couldn't speak, were poor hunters, and lived a brutish lifestyle that may have included cannibalism. Yet the evidence shows that they buried their dead, produced art, and cared for their elderly. It may be that we are just species chauvinists, who can't bear the idea of others with bigger brains. We suggest showing some generosity of spirit, acknowledging the evidence that evolution produced several variants of humanity, with a range of brain sizes; and that we fall into the lower end of the recent scale of hominid brain size.

Perhaps Neanderthals are so reviled for another reason: the uncomfortable fact that their brains were bigger than our own. Neanderthal brains were in excess of 1,500 cc, more than 10% bigger than ours. Their big brain is often dismissed as a fluke. Yet their bodies were not appreciably larger than ours, making their big brains even larger in terms of brain to body proportion, and within-species brain variation (e.g., a human jockey vs. basketball player) is small, adding significance to the large Neanderthal brain. Scientists have for years been assembling a nearly-complete sequencing of the genes of Neanderthals, and fittingly, that work is nearing completion by a group of scientists in Germany, where Neanderthals were discovered. The genetic evidence shows substantial differences between Neanderthals and humans, suggesting that they may indeed have been separate species. Exactly when they disappeared is unknown. Some say they were still around 25,000 years ago, possibly carrying on an active trading business with modern humans. How long might a memory of these big-headed, immensely strong near-humans reverberate down through generations of storytellers? The Bible tells us that "there were giants in the earth in those days . . ." And the first recorded epic, Gilgamesh, describes a terrifying half-human monster, living alone, deep in the forest.

Thoroughly modern humans first show up in the fossil record about 200,000 years ago, with brains averaging around 1,350 cc. These were probably not descended from the *heidelbergensis* off-shoot, but rather are likely to be a separate branch emanating from *Homo erectus*. These humans, *Homo sapiens*, spread rapidly across Europe and Asia, eventually reaching the far northeast corner of Siberia, from which they spread into the Americas.

If we put all these skull sizes together, a striking picture emerges. Figure 11.3 shows the skull volumes for each of the species noted above, according to the time they appear in the fossil record.

We immediately see two discontinuities in the graph. Brain growth seemed to stay stable through two million years of Australopithecines, then took a jump to *Homo erectus*, stayed stable again for perhaps a million years, and then jumped again with the arrival of the first archaic *Homo sapiens*. This graph shows absolute brain sizes; if we were instead to show relative brain to body ratios, the effect would be the same, with largely flat periods interrupted by two distinct jumps. These jumps provide clues to our history.

Apparently, there were two remarkable events in the history of the hominids, one about two million years ago (the appearance of *Homo habilis* and *erectus*) and the other just 500,000 years ago (archaic *Homo sapiens*). What could have happened at those time points to cause such a radical and sudden expansion of the brain? This constitutes one of the greatest enigmas in the origins of human beings.

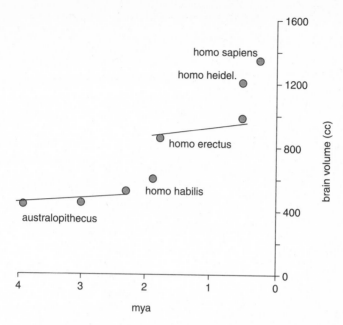

Figure 11.3 Brain sizes in the hominids. Australopithecines (beginning four million years ago, or mya) had brains sized much like those of present-day apes. An apparently abrupt jump in brain size occurred about 2 million years ago with the arrival of *Homo habilis* and early *Homo erectus*. Brain size remained stable for roughly a million and a half years. Then another jump occurred barely half a million years ago.

A popular explanation is often heard. It derives from the fact that our early ancestors, just recently divorced from the apes, moved from the trees to the open savannah. Perhaps we there encountered new and terrible types of competition, which may have led to pressure for novel evolutionary adaptations. Perhaps then our hands, freed from the ground by walking upright, began to be used for carrying objects, then for fashioning tools. These activities may have demanded great brainpower, giving competitive advantage to individuals with larger brain areas for hands and tools. These competitive pressures would have had to explode around two million years ago, increasing brain size and perhaps generating new social behaviors for defense and for predation.

According to that hypothesis, the brain grew large because it needed to pack in just those features we happen to value: tool use, intellectual leaps, social interaction. An almost irresistible notion, and again possibly the irresistible fallacy. By this concept, brain expansion is driven by specialized needs, with add-ons growing in mosaic fashion.

But many scientists take issue with such explanations. We have just seen that the components of the brain change with great predictability. It is difficult to support the notion that external pressures caused all of these brain component size changes individually. Evolutionary biologists Barbara Finlay and Richard Darlington, among others, have published an array of studies showing results like those of figures 11.1 and 11.2. The brain doesn't grow mosaically, differentially adapting itself to external pressures. It grows in uniform, concerted fashion, according to its own internal rules; and whatever behavioral abilities happen to emerge from it are side effects. Finlay and Darlington have shown that as a brain develops in an embryo, vast tracts of cortex are created according to principles having to do simply with overall brain size, and little else. The brain grows huge, and its parts become secondarily exploited for behaviors.

These two hypotheses, mosaic and concerted evolution, make different predictions that can be examined. Mosaic evolution predicts that the different parts of a brain will differ substantially in different species: the relative size of different parts will be different in humans, chimps, and apes, because, as per the theory, these differences will be forged in response to environmental pressures. For instance, the zones responsible for planning and for speech might be prominent in one case and not in the other, while areas for sensory processing might be expected to be similar across groups. But the proportions of the human brain are exactly what would be expected for any primate brain of its size, as predicted by concerted evolution. That is, once the brain has grown to 1,350 cc, then its parts are just as predicted for that size brain (figure 11.2). Similarly, the relative sizes of other structures such as hippocampus and striatum are in no way exceptional. The retained relative proportions of all these brain structures is strongly supportive of concerted evolution. If evolution was acting externally on the brain, as the mosaic theory proposes, it must have coincidentally left the brain's proportions unchanged. And this is highly unlikely to be the case.

As an example of the mosaic argument, it has sometimes been argued that the newest parts of the brain, the frontal lobes, are unexpectedly large in humans. But they are no larger than they are expected to be in a 1,350 cc brain, as predicted by concerted evolution. In particular, Finlay and Darlington demonstrate that those regions of the brain that arise later during infant development are the ones that grow the largest as overall brain size increases; the rule is often referred to as "late equals large." Another example has to do with language, and lateralization: the differences between the left and right sides of the brain. Humans have a language-critical zone, Broca's area, on the left side of the frontal area but not on the right; the mosaic argument might suggest this to be a language-specific change, occurring only in humans. But recent work demonstrates that a chimp brain is similarly lateralized, though correspondingly smaller, as expected by concerted evolution. No differences appear that are inconsistent with the concerted evolution of the brain—its expected proportions stay put as it grows larger.

*   *   *

Human brains do have some features that distinguish them from other primate brains, including subtle differences in these language areas, which we'll discuss more in chapter 13. We have been separate from other primates for five million years, during which small random genetic changes have likely taken effect. Even different patterns of development and diet are likely to affect us; and the most-affected structures will be those that mature most slowly during development, such as the brain. Imagine that a car company spins off a foreign subdivision, and 30 years later we take a look at the vehicles they're each constructing. Even if both had held to the original corporate philosophy, the cars will likely differ in hundreds of details. Perhaps one company evolved a certain type of interior, while the other developed a predilection for chrome wheel covers. But all car designs have significant constraints: they need to accommodate drivers, to have some reasonable fuel efficiency, to be able to accelerate and brake in adequate amounts of time, so strong similarities between the different cars will likely remain.

The primate brain is similarly regulated by basic constraints, in this case genetic and developmental ones; these initiate relative differences in the size of brain parts, early in gestation, and the differences are then magnified proportionally by the speed and duration of pre- and post-natal growth, as proposed by Finlay and Darlington. The contributions of selection pressures are certainly present; some mosaic evolution occurs. But these differences are minor in comparison to the powerful influence of the intact genetic plan, which specifies the brain's proportions as per concerted evolution.

BIG BABIES

Human brains, then, are qualitatively the same as those of other primates; the primary difference is that they are enormous. We've argued that the relative parts didn't change; just grew. And we've shown that the irresistible fallacy doesn't give an answer: it wasn't external pressure for intelligence that grew our brains, any more than external pressure for stupidity shrunk them back from the size of the Neanderthals and Boskops. Something caused great brain increases twice: two million years ago, and half a million years ago. What was it?

The fossil record doesn't show the fine details of biological processing, so we turn to the living primates, looking for clues to brain expansion. There are two examples of brain size jumps in today's primates: from the prosimians (today's lemurs) to the monkeys, and then again from apes to humans. Is there anything in common in these jumps, any clue to their origin? In both cases, there were big increases in the size of the newborn baby. Let's examine, then, the baby sizes that occur when the brain jumps occur: first in the move from Australopithecus to *Homo*, and then from *Homo habilis* to *Homo sapiens*.

We humans have huge babies. At birth, we are twice as large as a chimp baby, and 60 percent larger than a gorilla baby. It is notable that there is a severe cost for this: humans often die in childbirth. Indeed, childbirth mortality is far higher for humans than for other animals; this curse has accompanied us from the beginning. (In Genesis, God singles out Eve: "I will greatly increase your pain in childbirth; in pain you will bring forth children.")

What do big babies have to do with big brains? It turns out that one strongly predicts the other: just knowing the size of the newborn tells you almost all you need to know about the size of the eventual brain that will grow. Could we have developed big babies, and gotten big brains as a result? The irresistible fallacy immediately suggests the reverse relationship: perhaps we evolved to have big babies precisely so that we could grow big brains. Maybe the pressure for intelligence somehow gave rise to gene changes that expanded the uterus during pregnancy. But, resisting the fallacy, we note that lumbar and pelvic regions are most strongly influenced by locomotor adaptations: that is, it's most likely that genetic changes arose in how we walked, and that big brains arose as a side effect.

*   *   *

The hypothesis, then, says that big hominid babies appeared with *Homo habilis* and *erectus*, and then again with the archaic *Homo sapiens*. Knuckle walking, practiced by chimps and especially gorillas, comes with a rigid lower back. To achieve this, the number of vertebra in the spine were reduced in the apes, foreshortening the lower spinal cord. Then, when our ancestors began to walk upright, the spinal cord changed from a rigid beam, carrying the body, into a vertical column; and then that column is bent to move the shoulders into balance over the top of the hips, rather than out in front. Taken together, these arrangements are a recipe for lower back pain, but they do an excellent job at constructing a frame that can walk and run upright. They're apparent in a glance at human and chimp bodies: the lumbar (lower back) region of a chimp accounts for only about 20 percent of his overall trunk length; while the same region of a human measures almost twice as much, almost 40 percent of trunk length. Evolutionary changes that let us walk increased our lower trunk. As a side effect, our lengthened lower trunk makes an extended space into which the uterus can expand during pregnancy. The first brain size jump occurred when *Homo habilis* and *erectus* split from Australopithecus. They began to walk upright, erasing a constraint that had kept the size of the fetus small.

The second jump in brain size, 500,000 years ago, was not accompanied by obvious changes to walking. Our ancestors already had walking down pat. But there were apparent adaptations to childbearing. The pelvic girdle of female humans is noticeably different from that of males, perhaps in response to difficulties of being pregnant while walking upright. Much debate surrounds the origins of those gender differences, but the end result is the same: a larger pelvis enables the development and delivery of still larger babies. And as we've seen, larger babies grow larger brains.

We propose, then, that the big brain arose from the big baby, and the big baby arose first from changes in walking, and then in enlarged hips in females. The proposal thoroughly contradicts the alternative assumption, that selection pressures for intelligence drove the evolution of big brains. Tool use didn't prompt brain expansion. Rather, walking expanded brain size, and the bigger brain was able to conceive of tool construction and use.

This hypothesis seems destined to be unpopular. It says that chance dictated the essential event in the multi-million year history of humanity. The apes chose a particular way of walking, which had the side effect of restricting newborn size, and thus closed off the possibility of their brains expanding. Then early *Homo* species evolved a different way of walking, which happened to remove the restriction on fetal growth. This in turn happened to produce bigger brains, and higher intelligence, than had ever before been seen on the planet. When we're species chauvinists, we tend to see ourselves as the pinnacle of creation. When we resist the chauvinistic fallacy, we find that a plethora of otherwise confusing evolutionary data, on bipedalism, baby size, tool use, and sexual dimorphism, now fits together cogently.

ON INTELLIGENCE

Bio-mechanical adaptations, first to walking and then to the birth process, produced two rather sudden increases in the size of the newborn. The standard primate rules linking newborn size to brain size inexorably generated the big brain. Little role is left for selection of expanded brain regions that happen to execute human mental operations. In general, the role of intelligence *as a driver* of human evolution has, in our estimation, been greatly exaggerated.

This in no way denigrates the dominating role of intelligence in the history of the humans. Once the brain grew, and with it our intelligence, we began to be freed from the physical demands of the environment. Intelligence allowed *Homo erectus* to build fires for heat in the winters of Europe, to light caves in the summers of sub-Saharan Africa, to hunt in the forests of Indonesia and the hill country of Asia, all without marked biological adaptations to local weather.

If intelligence were a series of specialized responses to particular external pressures, it would be an amalgam of separate abilities. But intelligence from unpressured brain growth is different. Features that we consider to be uniquely human are actually hiding, albeit in an undeveloped, primitive form, within the ape brain. Earlier we noted that researchers have identified a lateralized zone in the chimp frontal lobe that they believe to be the precursor to the speech zone. They speculate that right-handed chimps use this left-brain area to make communication gestures or signals. As these areas expand in our big brain, then experience-induced modifications allowing access to the mouth and throat provide a substrate for cortical control of speech. As the big brain spurts vast tracts of association cortex, those repeated computational circuits used their newfound skyrocketing capacity to build ever-larger versions of the same structures. There were no shockingly new circuit types—just more of the same, in outlandish excess. What possible good does this do?

For one, big brains acquire a truly immense capacity for storing arbitrary information.

Professor Leo Standing, a Canadian psychologist, once set out to test the capacity of this memory. He recruited a class of psychology students to view 100 pictures, each presented for just five seconds. Then he brought them back in a week, and showed them those pictures again, mixed with 100 new pictures, and the students were asked to tell the researcher to push a button if they've seen the picture before. The students correctly recognized more than 90 of the pictures, having seen them only once, for just five seconds. It was a challenge to Professor Standing: how many images would he have to show before the students began to forget some? He ran the whole experiment again, this time with 1,000 pictures. Again, each shown for five seconds, then tested the students days later. Amazingly, the students again achieved better than 90 percent correct recognition. So Professor Standing decided to go the distance. Not with 2,000, or 5,000, but with 10,000 images. Again, five seconds per image. And again, unbelievably, better than 90 percent recognition. Professor Standing gave up, and published a paper on his findings, called "Learning 10,000 Pictures." We have no idea what would happen if we tested with 100,000, or 1,000,000, but one suspects that the results might be similar. The capacity of memory seems mind-numbingly, almost impractically huge, beyond anything we can imagine using it for.

If we lived for 100 years, and memorized a new picture once every minute, for twelve hours a day, we'd see a bit more than 25 million pictures. It's possible that we'd succeed in remembering them all: that our memory capacity is large enough to keep storing new information, just about continuously for a lifetime, or more. An overabundance of capacity.

In the frontal association areas, the story is the same, and more: enormous expansion of capacity, along with huge, thick cable bundles connecting brain areas to each other. One side effect of these bundles is the ability to store longer and longer sequences. In a normal sized brain, like that of a horse or a monkey, memory may be like a scrapbook collection of snapshots. In our brains, the extra capacity is so excessive that the pictures run together to be nearly continuous. The resulting "episodic" memory is prevalent in us: we think back to events and can reconstruct them almost like a movie. (Intense research is exploring whether similar effects occur in other animals.) Using our vast memories, we can call up long sequences, rearrange them, add to them. We can also recall them almost as if they were happening in the present time: we reactivate sensory images so strongly that we can practically hallucinate them.

This ability may be at the heart of human intelligence: the capacity to take a series of past experiences, and manipulate them to produce different outcomes. From our everyday ability to plan ahead, to the remarkable abilities of some to picture complex outcomes, we rely on vast communication channels connecting enormous networks all of the same type. These long episodes have a striking characteristic: we weave together sequences from our vantage point, as though an unseen entity were holding an unseen camera. We designate that unseen observer "me," floating behind our continuous memories. We see this market, that Parisian boulevard, this train station, all from the perspective of an entity, an observer, a viewpoint. This too may be a central aspect of the unique memory capability of our big brains. It's likely that the same ability resides in the even bigger brains that came before us.

CHAPTER 12

GIANT BRAINS

We return to the scenario that opened this book. In the autumn of 1913, two farmers were arguing about skull fragments they'd uncovered digging a drainage ditch. The place was Boskop, a small town about 200 miles inland from the east coast of South Africa. These Afrikaner farmers, to their lasting credit, had the presence of mind to notice that there was something distinctly odd about the bones. As we've related, they brought the find to Frederick FitzSimons at the Port Elizabeth Museum, and thus the first Boskop skull came to light.

The scientific community of South Africa was small, and before long the skull was brought to the attention of Dr. S. H. Haughton, one of the country's few formally trained paleontologists. He reported his findings at a 1915 meeting of the Royal Society of South Africa: "The cranial capacity must have been very large," and "calculation by the method of Broca gives a minimum figure of 1832 cc." Boskop, it would seem, possessed a brain perhaps 25 percent or more greater than our own. The idea that giant-brained people were not so long ago walking the dusty plains of South Africa was sufficiently shocking to draw in the luminaries back in England. Two of the most prominent anatomists of the day, both experts in the reconstruction of skulls, weighed in with opinions generally supportive of Haughton's conclusions. Sir Arthur Keith, president of the Royal Anthropological Institute, decided that Boskop "outrivals in brain volume any people of Europe, ancient or modern. . . ." Even more telling was the analysis by Grafton Elliot-Smith, easily the most talented English neuroanatomist of his time. Smith, a pioneer in the study of brain evolution, used a plaster cast of the skull's interior to estimate the brain's volume at about 1,900 cc. Boskop had arrived.

And the story back in Africa was still advancing. The Scottish scientist Robert Broom went back over Haughton's work. After carrying out measurements on a new endocranial cast of the inside of the skull, he reported that ". . . we get for the corrected cranial capacity of the Boskop skull the very remarkable figure of 1980 cc." Remarkable indeed: these measures say that the distance from Boskop to humans is greater than the distance between humans and their *Homo erectus* predecessors. In the previous chapter, we saw the great leaps in brain size during the past two million years; but the huge Boskop brains mean that these incredible surges in brain size didn't stop after the appearance of *Homo sapiens.*

THE MAN OF THE FUTURE

Might the very large Boskop skull be an aberration? Might it be caused by hydrocephalus, or some other disease? Broom and Elliot-Smith were both certain that the skull showed no such pathology, but the question was quickly preempted by new discoveries of more of these skulls. Mr. FitzSimons had kept busy, and had found another dig site, located in a rock shelter that apparently had been used in prehistoric times. Excavating some 15 feet deep into this shelter, FitzSimons found bones of "entirely different caliber and appearance" than the material above it. He sent some of these bones to the South African researcher Raymond Dart, who agreed that these were more instances of Boskops.

This new find contained even more complete skulls and skeletons. One such fossil had apparently belonged to a slender, 5'6" tall female. Her slight skeleton was yet again topped by a gargantuan head, with a brain volume that Dart estimated at 1,750 cc. He waxed eloquent on this find, noting that the brain was even greater than that of the Italian Renaissance painter Raphael, who supposedly had one of the largest human brains on record. Dart also commented that the other, less-complete skeletons and skulls appeared to have still greater capacity.

As if the Boskop story was not already strange enough, the accumulation of additional remains revealed an additional bizarre feature: these people had small, childlike faces. Physical anthropologists use the term "pedomorphosis" to describe the retention of juvenile features into adulthood. This phenomenon is sometimes used to explain rapid evolutionary changes. For example, certain amphibians retain fishlike gills even when fully mature and past their water-born period. Humans are said by some to be pedomorphic compared to other primates: our facial structure bears some resemblance to that of an immature ape. Boskop's appearance may be described in terms of this trait. A typical current European adult, for instance, has a face that takes up roughly 1/3 of their overall cranium size. Boskop has a face that takes up only about 1/5 of their cranium size, closer to the proportions of a child. Examination of individual bones confirmed that the nose, cheeks, and jaw were all childlike.

The combination of a large cranium and immature face would look decidedly unusual to modern eyes. But not entirely unfamiliar; such faces peer out from the covers of countless science-fiction books and are often attached to "alien abductors" in movies. The naturalist Loren Eiseley made exactly this point in a lyrical and chilling passage from his popular book, *The Immense Journey*, describing a Boskop fossil:

> There's just one thing we haven't quite dared to mention. It's this, and you won't believe it. It's all happened already. Back there in the past, ten thousand years ago. The man of the future, with the big brain, the small teeth. . . . He lived in Africa. His brain was bigger than your brain. His face was straight and small, almost a child's face. When the skull is studied in projection and ratios computed, we find that these fossil South African folk, generally called "Boskop" or "Boskopoids" after the site of first

discovery, have the amazing cranium-to-face ratio of almost five to one. In Europeans it is about three to one. This figure is a marked indication of the degree to which face size had been "modernized" and subordinated to brain growth.

And this:

> I have stared so much at death that I can recognize the lingering person-
> alities in the faces of the skulls.... One such skull lies in the lockers of a
> great metropolitan museum. It is labeled simply: Strandloper, South
> Africa. I have never looked longer into any human face than I have upon
> the features of that skull. I come there often, drawn in spite of myself. It is
> a face that would lend reality to the fantastic tales of our childhood.
> There is a hint of Wells' Time Machine Folk in it—those pathetic child-
> like people whom Wells pictures as haunting earth's autumnal cities in
> the far future of a dying planet. Yet this skull has not been spirited back to
> us through future eras by a Time Machine. It is a thing, instead, of the
> millennial past. It is a caricature of modern man, not by reason of its
> primitiveness, but, startlingly, because of a modernity outstretching his
> own. It constitutes, in fact, a mysterious prophecy and warning. For at the
> very moment in which students of humanity have been sketching their
> concept of the man of the future, that being has already come, and lived,
> and passed away.

Boskops, then, were much talked and written about, by many of the most prominent figures in the fields of paleontology and anthropology. Yet today, although Neanderthals and *Homo erectus* are widely known, Boskops are almost entirely forgotten. Some of our ancestors are clearly inferior to us, with smaller brains and ape-like countenances. They're easy to make fun of and easy to accept as our precursors. In contrast, we've pointed out that the very fact of an ancient ancestor like Boskop, who appears un-ape-like, and in fact in most ways seems to have superior characteristics to ourselves, was destined never to be popular.

But the very timing of Boskop's discovery played a role in its eventual obscurity. Boskop was discovered at the same moment as that another skull, this one in England, and also ballyhooed by many of the same scientists. The irony, for Boskop's prospects of lasting fame, is that this other skull, the Piltdown Man, was a complete fraud.

HOW GIANT BRAINS WERE FORGOTTEN

As we have pointed out, Darwin, along with most scientists of his day, got the story of hominid evolution backwards; he assumed that the brain evolved first, followed by an upright body. After all, went the reasoning, what point is there in walking upright, freeing up the hands for tools, if there isn't sufficient brainpower for inventing or using tools?

These arguments had considerable force, and likely shaped the expectations of what the fossils of early humans ought to look like. On the way from ape to human, Darwin predicted, there should be a stage with ape-like bodies topped by larger-than-ape brains.

In 1912, about fifty years after Darwin's prediction, just such a fossil turned up—and, incredibly, it was found barely fifty miles from his home, in England.

It was uncovered by workmen in a gravel pit in a region called Piltdown in southeast England, and brought to the attention of an amateur archeologist named Charles Dawson, a well-respected citizen who practiced law in the area. The fossil skull had a human-looking braincase paired with the receding jaw of an ape. Just as predicted, this brain apparently evolved before the body.

That such a fossil, initially termed "the most instructive and important of all human documents yet discovered in Europe," should be excavated within a brief train ride of the center of Imperial Science seemed too good to be true. And in fact it was: Piltdown was a hoax, and a carefully planned one. The discovered "fossil" was in reality an unusually thick human skull, paired with the jaw of an orangutan, both broken in such a way as to remove all zones of interaction between the two and then chemically doctored to look ancient. The hoaxer had even carefully filed certain of the teeth on the orangutan jaw to give the appearance that they lined up with the (human) upper jaw. The pit was also salted with genuine fossils of primitive elephants, as well as a collection of chipped stones like ancient hand axes.

Dawson took his find to his friend Arthur Smith Woodward, a prominent paleontologist at the British Museum, who then assisted in further excavations. More bones and teeth were slowly uncovered. Sir Arthur Keith tells us that rumors of the great discovery were circulating in London scientific circles in the late summer of 1912, and that there was a packed house for the formal presentation given later that year at the Geological Society meetings.

> A great company assembled in the rooms of the Geological Society of London on the evening of December 18th, 1912, to receive the first authentic account of the discovery at Piltdown. An unknown phase in the early history of humanity was to be revealed; a revelation of that kind stirs the interest of many men, and draws them from their studies and laboratories to brave the heated atmosphere of overheated meeting-rooms. . . . It was quite plain to all assembled that the skull thus reconstructed by Dr Smith Woodward was a strange blend of man and ape. At last, it seemed, the missing form—the link which early followers of Darwin had searched for—had really been discovered.

Soon all the luminaries of British paleontology were drawn into the Piltdown story. The bones were examined minutely, errors in reconstruction of the skull were corrected, and the story grew. A very impressive assembly of scientific talent agreed that the skull and jaw belonged to the same individual.

Outside of England, doubts appeared rapidly. Deep skepticism was expressed by the very talented Czech-born American paleontologist Ales Hrdlicka, from his position at the U.S. National Museum, the nascent Smithsonian Institute. In a 1913 paper, Hrdlicka argued that Piltdown's skull and jaw could not have come from the same creature, or for that matter from members of the same species. Hrdlicka also was engaged in professional battles over the site of origin of the genus *Homo*, and he may have been particularly irritated to abruptly see England now proclaimed as the ancestral home site. Other American workers also apparently quickly recognized that Piltdown was a hoax. And it clearly was not just Americans who opposed the British consensus; the Frenchman Marcellin Boule, a world authority on prehistoric tools, also weighed in with a strong negative opinion, as did several German scientists. Looking at the list of believers vs. skeptics, it is hard to escape the conclusion that it was the British establishment against much of the rest of the world; quite possibly internal pressures and academic ambitions clouded the judgment of those closest to the fossils. The Piltdown episode is sometimes dismissed as the result of the primitive condition of science at that time, but perhaps this is too facile. In light of the separation of believers in England and doubters most everywhere else, it seems that there is more to the story: sociological factors perhaps involving personal and national pride. The Piltdown episode is, in one sense, simply an instance of outright fraud, something that is very rare in science—but the sociological influences that shaped the professional reactions to Piltdown are, unfortunately, far more endemic.

We cannot leave this history of the Boskops without some discussion of who perpetrated the great Piltdown hoax. The list of suspects is extraordinary. The obvious villain, Charles Dawson, had motive—he desperately wanted to gain admission to the Royal Society—but is questionable with regard to means: the hoaxer must have had access to a large collection of genuine fossils and in addition possessed the knowledge to chemically modify bones so as to have made them appear to be fossils. And much more: the hoaxer had to break the bones and file the teeth so as to make an orangutan jaw line up with a human skull (on this point, he failed to convince the continentals and colonials). Many detectives find this to be too much for a simple country lawyer, and so conclude that Dawson was either a dupe or acted in league with a scientist. A big splash came with a very engaging book ('The Piltdown Men') claiming that the hoaxer was Grafton Elliot Smith, a formidable scientist who certainly had the means and, according to the book, felt slighted by the establishment because of being an Australian. The author also raised a point that continues to fascinate: the hoaxer could have assumed that the bunglers within the establishment, after buying the Piltdown story, would be massively embarrassed when more competent investigators denounced the whole thing as preposterous. The image of the hoaxer gnashing his teeth while the subjects of his cruel joke blithely use it to accrue ever-greater glory is indeed appealing. One detective after meticulously sifting through the evidence decided that the mischief-maker was none other than Sir Arthur Keith, the doyen of English paleontologists. The evidence is circumstantial—Keith seems to have known a great deal about the finds before they were made public and had surreptitious contacts with Dawson—but at the least is intriguing. And then there is Arthur Conan Doyle. The beloved creator of Sherlock Holmes lived a few miles from Piltdown, regularly visited the site, traveled enough to accumulate the bones, and had a score to settle. Doyle fervently believed it possible to communicate with the dead via an appropriate medium. Certain members of the Piltdown cast had earlier triumphantly unmasked one of these mediums as a fraud and used this case to ridicule spiritualism and its followers. Doyle, so goes the argument, was bitter about this and wondered if the same scientists would, by the same logic, reject evolution if a fossil they embraced turned out to be a hoax. The problem here, of course, is why Doyle, after so brilliantly lining up his targets, didn't pull the trigger? Nearly as dramatic a possibility as Doyle is the Jesuit philosopher cum paleontologist, Pierre Teilhard de Chardin. (Of course! It was the Frenchman!) Teilhard had an extraordinary career. He not only made a major find at Piltdown, but years later participated in the discovery of Peking Man, one of the great paleontology digs of the twentieth century. Through all of this, and his work in prehistory, Teilhard compiled a theory about stages of existence and how consciousness would ultimately untether itself from space and time. The researchers who debunked the hoax in 1953 included him as a suspect in part because they thought him evasive in answering their questions. Louis Leaky, the discoverer of *Homo habilis*, and the noted evolutionary theorist Stephen Jay Gould, were both convinced that Teilhard was the man. A more recent investigator uncovered material that only deepens the mystery surrounding the Jesuit's actions during l'affaire Piltdown. Brian Gardiner, a professor of paleontology at King's College in London, has spent time and energy investigating the hoax. In his researches, he found a letter from Teilhard, written only a few days after the big Geological Society meeting, in which he comments that his scientific supervisor, the not-to-be-trifled-with Marcellin Boule, was not 'easily taken in', and so would be very suspicious about Piltdown. Teilhard, it seems, smelled a rat. Gardiner also points to a small paper published by Teilhard in 1920 ('Le cas d'homme Piltdown') that diagnoses the famous jaw as both belonging to a chimp and having been doctored. But realizing the bogus nature of Piltdown, why did Teilhard continue participating in the excavations? Unless, of course, he was the hoaxer. And finally, a noted anthropologist reported that Louis Leakey said that Teilhard had told him that he knew the identity of the hoaxer, and it wasn't one of the discoverers.

While all of the above are marvelous candidates, Brian Gardiner's research into the strange saga of Piltdown ultimately led to a smoking gun, and a far less romantic prankster. Martin Hinton was a volunteer worker at the British Museum. His day job was as a clerk at a law firm, but he spent his off hours diligently cataloguing fossil rodents. An old storage trunk belonging to Hinton was unearthed in the 1970s, and some of its contents were sent to Gardiner; these included a set of bones that had been chemically treated and structurally modified in much the same manner as Piltdown Man's skull and jaw. It emerged that Hinton's pet project had been to create an extensive catalogue of fossil rodents. He approached the overseer of paleontology at the museum, the very same Arthur Smith Woodward to whom Dawson would eventually bring the Piltdown fossils. Hinton asked Dawson for a weekly salary for his rodent fossil cataloguing, which Woodward refused, possibly earning Hinton's enmity. Hinton knew of Dawson's searches in Piltdown, and knew that Dawson would probably bring any finds to the senior scientist Woodward. Hinton may have decided to plant the doctored bones, and then to count on Dawson to act as Hinton's unwitting dupe. Perhaps the Piltdown story ends with those attic trunks, and the vengeful Hinton. But readers of mystery novels will recognize that the plot has too many suspects with means, access, and similar motives. They will be reminded that Agatha Christie's great detective Hercule Poirot was once faced with a case of this kind in *Murder on the Orient Express*. The Express was blocked by a snowstorm on its Paris to Istanbul run and, of course, a murder most foul occurred. Poirot, who happened to be aboard, found himself with too many suspects who matched, according to the impeccable logic that had solved so many earlier cases, the profile of the likely killer. Eventually, of course, his 'little grey cells' uncovered the astounding truth: each of the logical suspects had in fact committed the crime. It was a conspiracy. In Christie's story, the murderers had little in common but nonetheless had interacted with each other in the past. And so it was with Piltdown; Hinton, the unpaid worker at the museum, was introduced to the mystic Teilhard at Dawson's home. Arthur Conan Doyle, the already famous novelist, included several of the soon to be Piltdown men in his wildly successful The Lost World. Perhaps Teilhard, the lone excavator who knew how ridiculous the whole business was, had the responsibility of prolonging the game long enough until it crashed down around the heads of the British establishment. Could there have been an extended Piltdown Plot, involving many of these players? Where is that little Belgian detective when you need him?

To the British establishment, all other fossils paled next to the apparent brilliance of Piltdown. One casualty was Dart's great discovery of Australopithecus in 1924, which went largely unappreciated for decades. Dart submitted a paper on his discovery within weeks of making his find (an alacrity that was quite unlike normal practice at the time). The paper drew largely negative reviews; Keith, Smith Woodward, and others in the British establishment argued that Australopithecus, an upright ape-man with a small brain, did not fit their accepted pattern (advanced brain, ape-like facial features) set by Dawson's Piltdown. Some of the reviews were more personal and vitriolic, including an ad hoc challenge to Dart's competence with Latin. We can assume that it was similarly difficult to pay attention to the Boskop remains, the first human fossils from Africa. Boskop had very bad timing, arriving as it did within a year of the more celebrated Piltdown discovery.

As we've mentioned, there is more to the story of how the Boskops were lost to history. In chapter 1, we pointed out Robert Broom's prescient comments in the journal *Nature*, noting that Boskop "has an enormous brain and is not at all ape-like. Therefore, according to some, it cannot be old, and in any case cannot be very interesting."

The history of evolutionary studies has been dogged by the intuitively attractive, almost irresistible idea that the whole great process
leads to greater complexity, to animals that are more advanced than
their predecessors. The pre-Darwin theories of evolution were built
around this idea; in fact, Darwin's (and Wallace's) great and radical
contribution was to throw out the notion of "progress" and replace
it with selection from a set of random variations. But people do not
easily escape from the idea of progress. We're drawn to the idea that
we are the endpoint, the pinnacle not only of the hominids but of all
animal life.

Boskops argue otherwise. They say that humans with big brains,
and perhaps great intelligence, occupied a substantial piece of
southern Africa in the not very distant past, and that they eventually
gave way to smaller-brained, possibly less-advanced *Homo sapiens*;
that is, ourselves.

INSIDE THE GIANT BRAIN

We have seen reports of Boskop brain size ranging from 1,650 to
1,900 cc. Let's assume that an average Boskop brain was around
1,750 cc. This is 30 percent larger than our average present-day
1,350 cc human brains.

What does this mean in terms of function? How would a person
with such a brain differ from us?

Our brains are roughly 25 percent larger than those of the late
Homo erectus. We might say that the functional difference between
us and them is about the same as between ourselves and Boskops.

But there is no known way to take two different brains (especially
when one is long extinct), and estimate their different minds. We
have spent much of this book describing just what is in our brains,
and in the brains of other known animals such as apes, to enable
careful educated guesses about these potential differences. We
return now to these basic principles of the brain.

As we've seen, expanding the brain changes its internal proportions in highly predictable ways. From ape to human, the brain
grows about fourfold, but most of that increase occurs in the cortex,

not in more ancient structures. Moreover, even within the cortex, the areas that grow by far the most are the association areas, while cortical structures such as those controlling sensory and motor mechanisms stay unchanged.

Going from human to Boskop, these association zones will be even more disproportionately expanded. Careful anatomical work has been done on one of the keystones of all association regions, the most frontal part of the frontal cortex, a zone usually called prefrontal cortex.

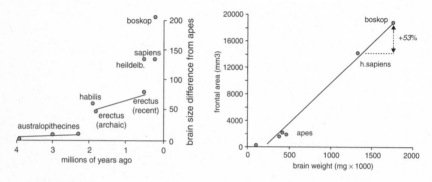

Figure 12.1 Relative brain sizes of the Boskops and other hominids. (Left) The graph shows how much bigger the brain is than the size found in an ape of the same body size. So Australopithecus, who was about the same body size as a chimp, had a brain that was barely 10% larger. Brain size abruptly increased by about 50% above the ape level around 2 million years ago in the small *Homo habilis* and the nearly human-sized *homo erectus* (also see figure 11.3). A further jump occurred with the advent of archaic *Homo sapiens*, adding about 150% more volume than would be found in an equivalent-sized ape. The Boskops carried the trend to an extreme by gaining more than 200% above what would be expected for an equivalent-size ape. (Right) The size of the frontal cortex (area 10) plotted against brain weight for the apes and modern humans. The value for the Boskops is extrapolated from the trend line. Increasing overall brain size from human 1350 to Boskop 1750 cc leads to an expected increase in frontal cortex of a remarkable 53%.

In figure 12.1, we add one last set of graphs to those presented in chapter 11. In the left-hand graph, we repeat the historical jumps in the record of our ancestors; these are the same data we showed in figure 11.3, but with two differences. This time we show hominid brains in terms of their relative differences from ape brains. And this time, we add a point for Boskops.

The first three low points are again Australopithecines. About two million years ago, a jump occurred with *Homo habilis* and *Homo erectus*. Then a half-million years ago, a jump occurred again, from *Homo erectus* to *Homo sapiens*. But then a new jump is seen: the leap from ourselves to the Boskops. As can be seen, this leap is as large, or larger, than the difference between ourselves and ancient *Homo erectus*.

On the right, we break out just the prefrontal cortex component of these brains. The apes and ourselves have expected frontal cortices for our overall brain size. We plot a point for Boskops according to the same principle. His brain size is about 30 percent larger than our own, that is, a 1,750 cc brain to our 1,350 cc's. And that leads to an increase in prefrontal cortex of a staggering 53 percent.

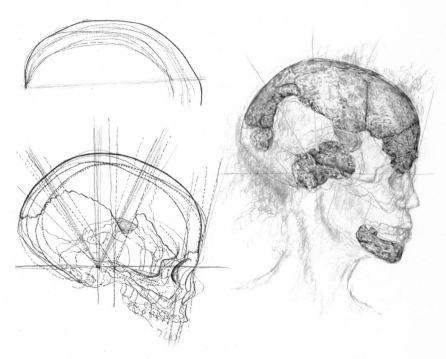

Figure 12.2   Possible appearance of the Boskops. Skull caps from fossils are shown at the top left. The light lines correspond to various Neanderthals, Cro-Magnons, and modern people. A Boskop skull is indicated by the dark line. More complete skulls are shown on the bottom left, with the Boskop again denoted by the dark line. On the right is a speculative reconstruction of a complete Boskop head. The shaded zones are drawings of particular bones uncovered by physical anthropologists.

So if these principled relations among brain parts hold true, then Boskops would not only have had an impressively large brain, but an inconceivably large prefrontal cortex.

The prefrontal cortex is closely linked to our highest cognitive functions. It makes sense out of the complex stream of events flowing into the brain; it places mental contents into appropriate sequences and hierarchies; and it plays a critical role in planning our future actions. Put simply, prefrontal cortex is at the heart of our most flexible and forward-looking thoughts.

The great expansion of prefrontal cortex from apes to humans is undoubtedly one of the major reasons why our behavior is so very different from theirs.

As discussed in earlier chapters, the representations built by these frontal association areas are hierarchical. That is, initial representations become constituents that are combined to build higher level representations. Representing a "school" combines representations of the constituents of a school—teachers, students, lecture halls—and each of those in turn is comprised of other constituents—background, degrees, admissions, and so on. Higher levels in the hierarchy naturally compress more and more information into each individual high-level representation.

Adding association brain areas adds more levels to the hierarchy, building richer high-level representations. This obtains for both high-level sensory memory and high-level motor memory. In the latter, each serial command in a high-level motor program has more detail; correspondingly richer episodes become possible.

While your own prefrontal area might link a sequence of visual material to form an episodic memory, the Boskop may have added additional material from sounds, smells, and so on. Where your memory of a walk down a Parisian street may include the mental visual image of the street vendor, the bistro, and the charming little church, the Boskop may also have had the music coming from the bistro, the conversations from other strollers, and the peculiar window over the door of the church. (Alas, if only the Boskop had had the chance to stroll a Parisian boulevard!)

Expansion of the association regions is accompanied by corresponding increases in the thickness of those great bundles of axons, the cable pathways, linking the front and back of the cortex. These not only process inputs but, in our larger brains, organize inputs into episodes. The Boskops may go further still. Just as a quantitative increase from apes to humans may have generated our qualitatively different language abilities, possibly the jump from ourselves to Boskops generated new, qualitatively different mental capacities. We internally activate many thoughts at once, but can only retrieve one at a time. Could the Boskop brain have achieved the ability to retrieve one memory while effortlessly processing others in the background, a split-screen effect enabling far more power of attention?

Each of us balances the world that is actually out there against our mind's own internally constructed version of it. Maintaining this balance is one of life's daily challenges. We bask in barely perceived attention, and rage at imagined slights. We occasionally act on our imagined view of the world, sometimes thoroughly startling those around us. "Why are you yelling at me? I wasn't angry with you— you only thought I was." Our big brains give us such powers of extrapolation that we extrapolate straight out of reality, into worlds that are possible, but never actually happened. Boskop's greater brains and extended internal representations may have made it easier for them to accurately predict and interpret the world; to match their internal representations with real external events. Perhaps, though, it also made them excessively internal and self-reflective. With their perhaps-astonishing insights, they may have become a species of dreamers, with an internal mental life literally beyond anything we can imagine.

GIANT BRAINS AND INTELLIGENCE

Dreamers perhaps, but very bright dreamers. Characterizing their overall intelligence brings us up against the limitations of our measures of human intelligence. The word "intelligence" itself is vague, and has a messy history. The notion of intelligence begins with the key assumption that there are unidimensional measures that characterize our abilities, and that these are correlated with each other. (Sometimes additional inferences are added: that these supposedly measurable traits are further correlated with overt traits such as skin color, or other imagined "racial" features, which we have shown do not actually even correlate with each other.)

As discussed, different individual abilities are separately exhibited, and often occur by themselves, uncorrelated to other characteristics. Scores on standardized intelligence quotient or IQ tests such as Stanford-Binet, Raven's progressive matrices, and the Wechsler scale, do correlate to a small degree. Some see this as evidence for a single underlying "g" or "general intelligence" factor. But a wide range of abilities can be measured by other, non-general tests, and these scores correlate poorly with each other and with "g." Even when correlation factors are found, they typically are small. For instance if a correlation coefficient of 0.4 is found between two variables (such as between an IQ test score and, say, job performance), that means roughly that just 15 percent of the variance (0.4^2) in job performance is accounted for by the IQ test score, and 85 percent is not! Clearly not a strong statement, nor one of much use in making predictions about how well a new employee will do.

(Hundreds of books and thousands of scientific articles have been written on "intelligence quotients" or IQs, and we will not even briefly review them here, except to point out a few undisputed facts: i) IQs are a single number, assigned to a person based on that person's scores on a small set of tests; ii) IQ test scores are normalized such that the "average" test-taker is intended to receive an IQ of 100; iii) scores are distributed via a statistical normal distribution, i.e., with most scores clustering around the mean of 100, and far fewer individuals receiving scores that are much higher or lower than the mean; iv) in some populations, IQ scores are correlated with behavioral and social variables including occupation and income; v) many individuals with high IQ scores perform very poorly at certain tasks, and many individuals with low IQ scores excel at certain tasks; vi) other testing systems yield separate scores on tests of different kinds of abilities such as musical, creative, memory, numerical.)

Rather, some people have unusual musical abilities; some, athletic abilities; some, abilities to build computer programs; some, advanced social facility. These tend not to go together. Scientists such as Harvard professor Howard Gardner have long formalized and studied these observations, demonstrating that there are scales on which people's abilities and capacities can be measured—and these multiple different measurement scales are independent of each other, not at all lending themselves to the notion of a single, one-directional, all-encompassing "intelligence" scale. As we have seen, these differential abilities could arise in part from different arrangements of brain paths. The diversity of abilities seen across individuals and populations may be partially due to different innate wiring, and partially to environmental influences on that wiring as the individual develops, or just chance differences in the way an immensely complicated cortex wires itself up.

Even if brain size accounts for just 10–20 percent of an IQ test score, it is possible to conjecture what kind of average scores would be made by a group of people with 30 percent larger brains. We can readily calculate that a population of people with a mean brain size of 1,750 cc would be expected to have an average IQ score of 149. This is a score that would be labeled at the genius level. And if there was normal variability among Boskops, as among the rest of us, then perhaps 15–20 percent of them would be expected to score over 180. In a classroom with 35 big-headed, baby-faced Boskop kids, you would likely encounter five or six with reported IQ scores at the upper range of what has ever been recorded in human history. And it could be even more extreme. Some researchers suggest that the size of the prefrontal cortex, not the overall brain, is the better predictor of IQ scores.

Lest this sound intimidating, we emphasize again that no one really knows what set of cognitive variables the IQ tests are measuring. It's likely that these tests are missing whole collections of intellectual capabilities. And much more work needs to be done before scientists can draw sensible inferences about the significance, if any, of differences in brain sizes in living people, especially when those measures are sensitive to what foods are eaten, hormone levels, and related variables. Though there may not be an IQ number to assign, it nonetheless may be concluded that Boskops would far exceed contemporary people in at least some mental capacities, including some that are associated in everyday parlance with being "smart."

CHAPTER 13

ALL BUT HUMAN

ON SCIENCE

The philosopher Willard V. Quine tellingly made the point that science is not proven; it is demonstrated. We can make observations, and the more densely connected those observations, the stronger the underlying facts seem to us. Does the sun come up every morning? It always has, but you can't *prove* that it will tomorrow; nor can any scientist. We can hypothesize a "fact," but we can never prove one, even a seemingly simple and obvious one. For extreme examples like that of the sun we have overabundant evidence, from basic physics, from astronomy, mechanics, not to mention sheer statistics: if you bet against it, you'd always lose. But other examples get tough in a hurry. Try instead to prove that dogs evolved from wolves, or that the drug Valium puts you to sleep by acting on GABAa receptors. In both cases, the facts are known as well as we know that the sun will come up—but it still is just evidence, not "proof." Now go farther: try to prove that birds evolved from dinosaurs, or that the drug clozapine helps schizophrenics by its dual action on dopamine and serotonin receptors. The evidence is convincingly strong, but now we're even farther from actual, experimentally demonstrable truth.

As Quine pointed out, we arrive at answers not by any magical method of "proving facts," but by carefully accreting experimental observations. Paradoxically, this scientific method works not because experiments prove facts, but because one by one, they rule out alternatives. The more we observe, the more we can dismiss some explanations that don't fit all the observations. What you're left with is the set of possible real facts. Sherlock Holmes said it well: "when you have excluded the impossible, then whatever remains, however improbable, must be the truth." Quine's web of observations constrains the possible facts ever more tightly, until almost all are ruled out, and the real explanation emerges.

When the pieces of evidence all fit tightly together in one of Quine's webs, we treat them as facts; when they don't, we discard them. But in between, there is a vast space of hypotheses, guesses, and predictions, and we have to keep on running experiments to see what observations really fit together. Science is the pursuit of knowledge at the frontier; along the boundaries between the known and the unknown. At one time, we didn't know of any relation between electricity and muscles; now we know in intimate detail how muscles use chemistry and electricity to send their signals. At one time, we didn't know how brain cells were connected, or how they send their signals, or how those signals are organized. Today we know an enormous amount, though there's still more to be learned.

Even well-known and tightly-connected facts can be shown to have odd exceptions. Isaac Newton and others had clearly worked out the details of mass, acceleration, and gravity, and they fit within a very tightly connected set of observations that could be used to accurately predict complex phenomena such as the movement of the planets in space. Newton's laws were the accepted "facts" of physics for more than two hundred years—but then, toward the end of the 1800s, an amazing sequence of scientists from Faraday to Boltzmann to Planck to Einstein began to note loose ends; observations that didn't fit. These eventually transformed Newtonian physics, spawning wholly new fields from quantum mechanics to relativity. Newton's laws still hold, for all the observations you can make in everyday experience; but the important exceptions turned out to hold for an enormous range of phenomena that occur outside our normal experience, explaining a far broader world.

Because of these triumphs, achieved over the course of centuries, physics can be used as an exemplar; in some ways the best of the sciences. The physicist Ernest Rutherford famously said "in science there is only physics; all the rest is stamp collecting." He was quite wrong, and not in the way he might have thought: just as other sciences depend on the collection of observations, so too does physics. We collect these experiments and observations in order to unite them in webs of tightly-knit theory. Physics does it, and chemistry, and biology; even psychology, economics, and sociology do the same. The difference is in extent, not in kind. The science fiction writer Isaac Asimov conceived of a future scientist, Hari Seldon, whose field was "psychohistory," capable of precise predictions about human affairs, as accurate as those for physics and chemistry. Where Rutherford was wrong, Asimov may have been right: such fields may some day come to exist, as observations and experiments in these far more complex fields achieve the level of predictive theory, rather than the early stage of "stamp collecting," or compiling observations, the wellspring of all science.

At any given time, the strength of facts are measured by their web of connected observations. To bolster hypotheses, we adduce multiple, connected observations consistent with them. To weaken them, we provide observations that clearly can't fit.

The further a theory gets from connected observations and explanations, the less we believe it. Much of the story of the brain can be followed all the way down to detailed biochemistry and subatomic physics. Built on those fundamental facts are many additional hypotheses of how the brain gives rise to the mind: how it stores information, relates observations to each other, and retrieves memories. The story of the brain proceeds all the way from physics to psychology; some theories are so tightly consistent that they would make Dr. Quine smile; others are still very much works in progress. In each case, the closer the findings are connected to already-strong observations, the more robustly they are supported, the more surely we can proceed.

We have, then, arrived at two hypotheses: humans acquire their powers primarily because their brains are big, or primarily because their brains are different from other brains. That is, as our brains grew huge, did they stay much the same, operating like an enormous, but otherwise typical, primate brain? Or did our brains also grow wholly new areas, new structures, new types of circuits, that might generate our unique powers? We note that these hypotheses are not mutually exclusive: some of our powers may come from size alone whereas others could come from new types of brain circuits.

Due to these hypotheses, the search is afoot for brain differences that could explain human differences. Scientists carefully examine human brains, and compare them against other primate brains, with special attention to those brain areas where we may be most different, such as the areas that are active when we listen to language, or when we ourselves speak.

DIFFERENCES

As we have mentioned, most structures throughout the thalamo-cortical system are remarkably similar to each other, and to corresponding structures in other primates. Differences have been extremely hard to find, and where they do exist, they are certainly far outweighed by the overwhelming similarities.

But those similarities among thalamo-cortical brain regions can themselves be outweighed by consideration of how different we humans *seem* to be. The cognitive gap between us and all other animals can appear an unbridgeable chasm. Possibly there are differences among brain areas, subtle ones, as yet undiscovered. The search for differences in thalamo-cortical structures of humans has been pursued with sedulous care. Scientists focus extra attention on brain areas that are in the anterior cortical areas, those that grew the most in the leap from ape to human, and in suspected language areas, in case they subserve new hidden faculties that could help explain the unique human language ability.

They're looking for particular kinds of things:

1. cells or circuits in the human brain that differ from each other, despite the rampant similarity among them—especially in locations that might be related to language or other very high-level cognitive abilities;
2. any cells, circuits or circuit designs that appear in humans but in no other animal;
3. circuits found in primates but are especially well-developed in humans;
4. any brain-related genes that occur uniquely in humans, or more in humans than primates, especially genes that may have come on the scene very recently;
5. overall shape changes in brains, which may suggest differential internal structure.

The idea is clear: to find design secrets that might contribute to an explanation of our differences. Machinery that could account for our unique abilities. Engines of the brain that could be driving language and reason. They're looking for the source of our humanity.

Differences *are* found. They are subtle, but perhaps they have explanatory power. The primary differences that have been found can be roughly divided into the above five categories.

Cell Types

In chapters 5 and 6, we pointed out that most cortical regions look the same. Throughout cortex, different areas have much the same neuronal cell types, arranged in the same layers. But a few slightly different cell types occur. In particular, in a few key brain areas, there are some deep-layer (layer 5) cells that are oddly elongated, like a pyramid whose bottom has been extruded straight downward.

These cells, named Von Economo neurons after their original nineteenth-century discoverer, occur only in the very largest-brained mammals. They occur in humans, as well as in gorillas, chimps and bonobos, but not in other primates such as orangutans, baboons, or monkeys. (Strangely, they also occur in large-brained cetaceans, including whales—which may provide further clues to these cells origins and their nature.) Moreover, these spindly neurons occur only in certain areas of our brains; specifically, the anterior cingulate cortex and the frontal insular cortex. In brain scans, these regions are active in situations of social interaction, especially when feelings such as trust, empathy, embarrassment, and guilt are triggered, suggesting that these odd cells may contribute to social cognition and behavior.

Local Circuits

The few exceptions such as Von Economo neurons prove the rule: most of the brain's array of neurons stays remarkably constant, across different animals and across their different brain areas. So too do the circuit designs into which these neurons are woven. The canonical circuits we have seen in chapters 5 and 6 are repeated endlessly across the cortex of humans and of other mammals, each nearly identical to the others. But just as there are a few neuronal cell types that stand out from their normal fellow neurons, there are circuits that show slight telltale differences from the canonical circuit layout. The differences are so slight that these slight outliers would never be noted but for the overwhelming regularity of design that characterizes all of their neighbors. A few brain areas exhibit a version of the canonical circuit that is slightly misshapen. In this version, the circuit is surrounded by extra material, rendering it cylindrically wider than the norm, while at the same time compressing the neurons within the column, packing them tighter together than they are in other areas. These "double-wides," measuring roughly 40 percent broader than typical canonical minicolums, appear to occur uniquely in brain regions in the left side of our brains, and specifically in left-side areas that are active in speech and language use: Broca's area in anterior cortex and Wernicke's area (specifically, a key part of it termed planum temporale) in posterior cortex.

Examination has shown what it is that makes up the expansion in these wider columnar structures. The extra material wreathing these wide columns is not composed of additional neurons, but additional wires: more axons than usual, running in and out, connecting these areas to other brain regions. Apparently these few areas make more connections than most.

Connectivity

All of these differences, from cells to circuits to connections, arise the same way—the only way they can—from changes to genes. It appears that our genes express much more of a particular protein that selectively initiate the formation of synapses. That protein, thrombospondin, operates in astrocytes, a type of glial cell that lives in the neuronal interstices. Glial cells do not communicate with other cells; do not send electrical messages like neurons—rather, they provide supporting functions to neurons, from growth to nutrition. Humans produce many times more thrombospondin than do either chimps or macaques, and that protein is found specifically in cortex, not in lower brain structures. The result is more connections being made in cortex—especially in the Broca's and Wernicke's language areas we've just seen, with their unique double-wide cortical columns, and possibly in many more regions.

These results seem to suggest a possible evolutionary specialization in cortical connections. Neurons in the human cortex have longer dendrites than those in an ape's cortex, and have substantially more synapses. Possibly, then, selection pressures for more elaborate circuits resulted in modifications to factors that control the thrombospondin gene. But brain size is determined by two factors: the number of neurons and the magnitude of their dendritic trees. So while the brain of *Homo sapiens* is three times as large as that of a chimp, it has nowhere near three times as many neurons. In general, the bigger the brain, the longer the neurons' dendrites, and naturally the more synapses per neuron. These increases place greater demands on local supporting (non-neural) glial cells, likely triggering greater activation of the thrombospondin gene. Thus the differences between humans and chimps are about as expected for animals with their brain sizes. This argument arises from the broader idea that brain development is a "decentralized" process. Genes lay down an early template, after which the growing pieces of the brain interact with each other via specific developmental rules. The longer this process proceeds, the more elaborate the final product. Features that are barely detectable in a quickly-assembled small brain can be greatly magnified in one with a longer developmental period. Those magnifications can seem like qualitatively new features.

Recent Genes

Any changes to cell types, circuit configuration, and connectivity, arise first from genes. Scientists can use "molecular clocks" to estimate how recently particular genes may have arisen in their current form. Surprisingly, they have been able to find genes that apparently have arisen in our genome in the very recent past.

- FOXP2 is a gene that is present in many mammals, from mice to men, but the version of the gene that we humans have is estimated to be less than 200,000 years old. People with FOXP2 mutations appear to have language-specific deficits, suggesting that the gene plays a role in our speech and language abilities.
- Microcephalin seems to have arisen within the past 50,000 years, and appears to regulate overall brain size. Evidence suggests that the current human versions of this gene were extremely rare, or nonexistent, until a few tens of thousands of years ago, but now occur in perhaps 70 percent of the entire human population.
- ASPM is a gene that occurs in cases of microcephaly, i.e., brains of reduced size. Evidence now suggests that the most recent version of this gene may have arisen within the last 10,000 years— a blink of an eye in evolutionary time.

Brain Shape

As we have seen, the internal connectivity pathways of the brain may be at least as important as its overall size. If the above genes enable brains to grow, we may ask what shapes they grow into. Studying the impressions inside a fossil skull can show the shape of the brain it contained, giving hints of which cortical brain areas might have been differentially enlarged. Some scientists have suggested highly differential brain growth, focused much more in some regions of forebrain than in others; in particular, a frontal brain region referred to as "area 10," widely thought to be active during complex reasoning and decision-making, is argued to have been enlarged far more than other areas. Again, we and others have provided substantial evidence that area 10 is about the size that would be expected for a 1350 cc brain (see chapters 7 and 12, as well as the notes for this chapter in the Appendix). Human brains, it is worth repeating, do have different proportions than those of chimps—differences which turn out to be those allometrically expected from the larger size of human brain.

These apparently disparate changes could all be related.

When new axons grow, they look for target neurons to connect to. If a particular target area, call it T, receives input connections from source areas A and B, what happens when those areas grow differentially? In a primate brain, areas A and B may be roughly equal, but as we've seen, in a larger brain, the later of these areas, say area B, will tend to grow disproportionately bigger. Now B's more numerous axons may take over the input to area T, simply overwhelming the input from A by sheer numbers. Even if A's inputs dominated T in a small brain, the growth of B in a big brain may generate a natural "invasion" of T from the profusion of new axons from the newly enlarged area B.

Thus brain pathways may get differentially lengthened or shortened, or diverted in new directions, due simply to the disproportionate growth that predictably occurs in brains of different sizes. Chapter 9 described the large brain pathways that define different assembly lines to which different types of perceptions and thoughts become assigned. Just as some brain areas grow disproportionately larger in big brains, so too do some brain pathways grow much larger.

Some brain paths, then, are longer than others, and some intersect extensively with others. Far more human brain is dedicated to visual and to auditory processing than to touch, or taste, or smell. And within our visual and auditory brain pathways, there are intensive assembly lines dedicated to some kinds of sights and sounds more than others: we process certain shapes more than we attend to color; we differentially listen to voices more than to screeches. The busiest processing "stations" in our brains arise from two anatomical types: path *endpoints* and path *intersections*. As processing proceeds down long brain paths, such as our visual and auditory paths, signals make their way toward endpoints, where extensive stretches of brain computation culminate, yielding our deepest hierarchies and most complex internal concepts. Similarly, when signals from one pathway intersect with others, such as the shapes of words interacting with their sounds, we generate new internal representations that cut across the boundaries of individual senses, combining to generate higher-level internal abstractions that transcend mere sights or sounds.

We have seen the example of the superior longitudinal fasciculus, the large human brain path that connects brain areas processing the sounds of words with the brain areas that process the visual shapes of words. Similarly, if our frontal area 10 is indeed disproportionately enlarged, it may be due to additional inputs, if area 10 turns out to be at a crucial endpoint or intersection among important connectivity pathways, which the evidence indeed suggests. And the areas we identify as uniquely conferring language abilities, such as Broca's area and planum temporale, may gain their apparently special status due to their privileged positions, at the ends and intersections of our lengthened visual and auditory pathways, which lead to them and connect them.

We again note that gene "control" of brain functions can be quite indirect. Gene variations have been linked to a number of cognitive disorders including Alzheimer's Disease and schizophrenia, involving memory loss and disordered thinking, respectively, but scientists don't then assume that the responsible genes are "controlling" learning or the proper sequencing of ideas. Rather, the gene variants typically (as in the case of Alzheimer's) encode proteins that regulate critical cell functions throughout the cortex, raising the question of why the disease so selectively impairs cognition, especially in its early stages. The answer likely has two parts: 1) more complex brain operations are likely to fail before simpler ones, and cognition is certainly complex; 2) cognitive problems are often more noticeable, whereas defects to more basic processes can sometimes be easier to hide. Schizophrenics, for instance, often have problems with rapidly repeated sounds, but these problems may be barely discernible to an observer. Clearly caution is in order when considering the relationship between particular mutations and particular cognitive functions.

FROM QUANTITY TO QUALITY

Perhaps the most lively (and sometimes vitriolic) debates in brain and cognitive science have to do with the question of where new abilities come from, and these often devolve to specific arguments about language. Put simply, how can humans have language, and how can no other living animals have it, given its immense utility?

In the 1950s, the linguist Noam Chomsky famously forwarded precisely that question, and proffered the attractive hypothesis that there must be special faculties underlying language: new language "modules" that human brains have and other brains do not.

Many theorists assume special faculties underlying language. We have just explored the current search for anatomically special structures that could uniquely reside in human brains. As we've seen, these studies have so far yielded exceptionally subtle differences among some brain areas, and no differences at all among many others. Scientists are still hard put to show any mechanism arising from these changes that could generate the vast differences seen in language. But the biggest change in brains is not in the cell types or wiring patterns, but in brain size itself. Moreover, we've seen that all of the differences are so far consistent with uniform changes arising just from the disproportionate growth that arises from brain size expansion, without resort to new structures that have special or unique explanatory capabilities.

We have described how the same internal constructs—sequences of categories, or grammars—can become dedicated to segregated "specialists" that uniquely engage different parts of the spectrum of perception and cognition. Different areas, such as those that specialize in recognition of faces vs. houses, are apparently constructed from nearly identical designs, and differ predominantly in the inputs they receive from areas earlier in the process. Their specialties arise from learning—from exposure to their particular inputs—which are selectively conveyed to them by their brain pathways, that is, by the structures they are connected to. Brain areas begin as largely similar, and get differentially "recruited" to a particular task. The combination of new areas growing disproportionately with brain size, and the recruitment of those areas to new tasks in these bigger brains, may directly explain the shockingly new behavioral capabilities of language.

There are many examples, albeit simpler, in which quantitative changes produce qualitative ones. As we pointed out in chapter 1, increasing the temperature of an 80° Celsius pot of water by one degree, and you simply have 81° Celsius water—but increase a 99° Celsius pot of water by one degree, and you have boiling water converting to steam. As long as uranium is kept below a critical mass, it will not detonate; above that threshold, it is explosive. Examples abound in physics, in chemistry, in biology, even in sociology: changes in size can yield differences in kind. And the transition from early hominid to human was no small one. Our brains are not 1 percent, or 30 percent, or 100 percent larger than those of comparably-sized chimps or apes, but roughly 400 percent larger. Intermediate brain sizes occurred in the early hominids; perhaps the transition to a language-ready brain occurred sometime during their occupancy of the earth. We may never know whether there was a moment of explosion; a time when the threshold was passed and we became human.

In contrast, some argue that simply passing a threshold is insufficient. Quality-from-quantity can be dismissed as a "Rubicon" argument, as if Caesar's crossing of that particular river actually made the decisive difference in his determined march on Rome. The counterargument is that there must be new kinds of designs that cropped up in the brain, and that these new circuits and systems must have had new powers that underlie the change from subhuman to human behavioral abilities. Such claims only add explanatory power when they propose mechanisms that could yield these new powers; these claims have their own dismissive epigram from critics: the "magic bullet" that somehow has a claimed effect (such as curing a disease), often through still-inscrutable means. If there are new circuits, they are still inscrutable, and the principles by which they might operate are of course unknown. Neither type of explanation is intrinsically invalid, but both are equally in need of demonstrable mechanisms that would enable them to work. Until such mechanisms are identified, both types of explanations are impossible to falsify, to demonstrate, or to rule out, and they may stay that way for some time to come.

FROM BRAIN ADVANCES TO COGNITIVE ADVANCES

There is a troubling gap in the brain evolution record that still cries out for explanation: our brains demonstrably were full human-sized (or even larger Boskop-size) by at least 100,000 years ago—yet the record does not show signs of cultural change, or uniquely human-like behavior, until much more recently; perhaps as recently as 20,000 years ago. This is more a problem for "threshold" explanations than for "new design" explanations. If a size threshold was crossed, it clearly was longer ago than the apparent rise of cultural changes; perhaps if a small, subtle new last-minute change was introduced to a few key brain circuits, this may have occurred more recently, and would have left no mark in the fossil record.

If you had been born in isolation—if no one had ever seen or spoken to you—you of course would not be able to speak or read. Nor would you know how to build a fire, or a home. You would seek shelter and warmth, but it might not occur to you to live in a house, or to wear clothing, let alone to actually create such artifacts. You would have the same brain and the same genes, but they would not have had a chance to express themselves in behavior.

Even with our large brains, we would likely be living much the way other primates live, if we all acted solely as individuals, and not as groups. To any given human individual, a big brain makes comparatively little difference. Its primary utility may well be to transmit information, especially information accreted over many generations. The "human-ness" of our lives depends on learning language as children, learning a variety of cultural characteristics ranging from clothing to shelter to vehicles to obtaining food and warmth, and then passing those on to our own offspring.

It is not surprising that our "human nature" is in part culturally determined. We are not born with intrinsic knowledge of houses and clothes, telephones and books. Our capabilities did not spring forth all at once, but coalesced incrementally over millennia. It is easy to pass culture forward once we have it, but it may have taken unknown spans of time to create it for the first time. Along the way, there may well have been "eureka" moments, such as rolling on wheels or the ignition and control of fire. These might have been accidentally stumbled upon at first, but then rapidly recognized as so useful that they were passed forward culturally.

If it indeed takes tens of thousands of years to accrete the trappings of civilization, then it may be no surprise that we humans lived on the planet for tens of thousands of years with the brain capable of doing so, before creating written texts, before building buildings, before inventing more than the most primitive tools. If the Boskops, with their enormous heads, tended to die in child-birth, they may never have built the accumulated, critical mass of society needed to truly capitalize on their intelligence.

FROM COGNITION TO LANGUAGE

If brain areas can begin as generalists, and become recruited to specialize in processing particular complex stimuli, then it remains highly possible for all brain areas to be constructed alike, and then to become specialized by connectivity and by repeated learning from experience.

Examples of this process have been convincingly demonstrated. A particular brain area, the "fusiform" region of the cortex, responds selectively when we see a face, rather than any other image. In adults, these regions are substantially larger than they are in children, growing roughly three times in size between late childhood and adulthood. Studies of this change have suggested that the face-responsive area in children is a subset, a central core, of the area that eventually expands into adult size. In the adult brain, the whole fusiform cortex responds selectively to faces. Researchers found that in the child's brain, only a small core region is selective to faces, but the remaining area is not selective. Instead, this surrounding sphere responds not just to faces but to many different visual objects: dogs, houses, spoons. In the adult, the entire fusiform cortex area, core and surrounding area, becomes recruited to specialize in faces. An area that began as a generalist, responding to many types of images, develops into a specialist, responding selectively to faces. The area is recruited to its specialized task, perhaps by the predominance of faces that we see in normal life, and by the connectivity pathways that no doubt run to these areas.

But fusiform cortex in all other ways seems just like every other area of cortex. As in language areas, a search goes on for any slight differences that could explain its specialized behavior, but it remains possible that it is not different: it truly is a general piece of cortex that becomes specialized to a task by dint of where it lies in the connection paths of the brain.

This is the larger question: what kind of machine is it that can give rise to new kinds of specialized behavior (e.g., language), just by making more of the same kind of machine? Most machines do not change the nature of their processing just by being increased in number, but cortex seems to grow new functions just by growing more cortex.

Grammars are, in this respect, highly unusual constructs. As we have seen, a grammar is a set of "rewrite rules," in which sequences of categories are rewritten to substitute category names with category members. Indeed, it is this ability that gives language its power and range: from a finite set of elements (words), grammars can create infinite numbers of strings (sequences) of those words—such as sentences. We can string new words together to make entirely new sentences under the sun, even meaningless ones, that nonetheless are readily recognized as grammatically sound. In the terminology of Chomsky and his colleagues, linguistic grammars are "generative," that is, they can be used to specify multiple instances of sentences without end. The generative properties of human language are what set it apart from other animal communication systems.

The question has often been framed in these terms: how do generative grammars arise in human brains? It is not obvious how these linguistic structures could arise from the presumably "lower" perceptual and motor processing in which other parts of the brain engage.

The hypotheses presented here turn that question on its head. As we posited in chapter 8, most brain processing, from perception onward, involves grammars, and the question instead is how the human brain extends these constructs from perceptual grammars to linguistic grammars.

Although the specialized "front end" circuits of the brain, with their point-to-point circuit designs, specialize in their own particular visual or auditory inputs, the rest of the brain converts these to random-access encodings in association areas throughout cortex. We showed in chapter 8 that these areas take initial sensory information and construct grammars. These are not grammars of linguistic elements; they are grammatical organizations (nested, hierarchical, sequences of categories) of percepts—visual, auditory, and other. Processing proceeds by incrementally assembling these constructs, and hierarchically passing them through long brain pathways, successively modifying them at each processing station in the path.

The grammar structures thus generated in the brain are used both in the processing of input, and in the generation of arbitrary new sequential outputs. In each case, they always operate strictly by the rules inherent in their hierarchically organized representational structures. In other words, these nested "sequence of category" structures from chapter 8 have the same generative property as other grammars, which are absent from many other mechanisms that are purely input-processing or statistical engines. From nested sequences of categories, a potentially infinite set of strings can readily be generated, and the resulting strings will be consistent with the internal grammar.

As we noted, these grammars generate successively larger "proto-grammatical fragments," eventually constituting full grammars. They thus are not built in the manner of most hand-made grammars; they are statistically assembled, to come to exhibit rule-like behavior, of the kind expected for linguistic grammars. Proto-grammatical fragments capture regularities that are empirically found to suffice both for recognizing and generating grammatical sequences. Research in our laboratories is in progress, studying the formal relations between typical linguistic grammars, and the proto-grammatical fragments that emerge from nested sequences of categories.

LEARNING CURVE

An additional characteristic of language that challenges researchers is the seeming effortlessness with which children learn language— readily contrasted with the comparatively laborious training typically required for adults learning a second language.

We humans learn their native language effortlessly; without formal training, without school, without books, without anyone telling us, without knowing that we're learning anything at all. From infancy we begin picking up sounds, and then words and phrases, and then structured human linguistic utterances in whatever language happens to be spoken around us. There is enormous debate about how this ease of learning occurs.

Other animals, to be sure, have communication systems. Chimps can be taught words and relations; so can dolphins; these are the equivalent of the communications seen in a 12- or 18-month-year-old. In this case, language learning is far from effortless; chimps must be taught laboriously and must be trained to attend and to practice, where humans simply soak language in. Nonetheless, it was once thought that if these initial levels could be achieved by nonhumans, then the next steps could somehow be reached. But despite enormous efforts over decades by researchers working with animals, no chimp, or ape, or dolphin, or any nonhuman creature, ever has attained the linguistic structures that all humans effortlessly achieve by age two or three.

Those who study language have long noted the distinction, and long argued that the vast gulf between human language and all other animal communication is a clear mark of a new faculty in human brains and minds—a set of structures as different from those in chimps, as hands are from tails. We observe the evolutionary creation of new body structures; why not new structures in the brain?

We have seen that the search for wholly new structures in the brain yields some tantalizing possibilities, from special cells to special genes, possibly underlying our language abilities. The similarities still far outweigh the differences among different areas in a human brain, but some presumably-innate, presumably uniquely-human tendencies enable children to master complex language structure solely by exposure rather than by intensive schooling. If an innate bias related to sequences of categories of vocal utterances (speech) led, in larger-brained organisms, to a downstream bias for certain sequences of categories of assembled speech sounds (words), then this may at least in part account for this much-studied but still elusive characteristic of innate language capability.

In other words, brain pathways evolved to lengthen the assembly lines dedicated to human sound sequences. Long assembly lines already existed, selected to process human communicative sounds that signaled emotions (contentedness, warning, fear, joy) and possibly became specialized to related content (predator, prey, different kinds of food, instructions to approach or maintain a distance).

Lengthening these lines may have sufficed to create far-downstream processing stations at which categories of utterances were input, and further constructs were pieced together from these.

Our hypothesis is a novel one. As we showed in chapter 8, thalamo-cortical circuits intrinsically create sequences and categories, and organize them into grammars. To apply this process to any given construct requires that an assembly line exists, which receives the inputs to be processed, and has plenty of ensuing room in the assembly line to create ever larger constructs from those elements. Auditory pathways in our brains grew and lengthened, building voice sounds into words, words into phrases, phrases into sentences.

It takes extensive processing, far down an assembly line, even to produce constructs corresponding to words, i.e., single sounds that denote objects and actions. Animals with intermediate brain sizes have sufficiently long brain paths to achieve this, enabling simple communication. To create complex sequences of words requires a preternaturally long brain connection pathway: long enough to arrive at words and their corresponding concepts, and to go beyond them. Whereas words are grammatically organized sequences of sounds, longer clauses and sentences are grammatically organized sequences of words.

The hypothesis begins with the operation of thalamo-cortical circuits, the computational analysis of their grammatical processing, and from the existence of brain pathways, to the differential growth of those pathways with increased brain size. From these multiple ingredients we forward this specific hypothesis of the origins of language.

FROM SPEAKING TO WRITING

We have seen that particular brain areas are differentially connected into long and strong assembly lines in talented readers, and that those lines are weaker or more diffuse in those with fewer reading skills. This represents an even further assembly line, and one that apparently has been recruited culturally rather than evolutionarily. Reading is not innate, and it is not effortless (unlike learning of spoken language). Because it requires laborious effort, much of early schooling is dedicated to reading.

As we saw in chapter 9, reading also is a marker that separates people with slightly different brain pathways; better readers have different brain pathways than poorer readers. Reading is a task that is so difficult that it is an aspect of the brain where differences are perhaps most likely to show up. Whereas innate tasks are almost universally acquired by all people, non-innate tasks must recruit brain areas that were not specifically evolved to perform them, and thus will fail in some percentage of people.

Reading involves: (1) understanding spoken language; (2) mapping individual spoken speech sounds (letters, not just words) on to written marks; and (3) recognizing sequences or assemblies (in pictograms) of those marks as constructs that map to words. Thus a full assembly line for taking speech to language must be present, and its processed outputs must be conveyed to an assembly line that performs the sound-to-visual (letters) mapping, and then that new assembly line must be sufficiently long to construct the written words in the language.

This may be taken as suggestive evidence for a final stage that led to very recent human evolution. When systems of writing were invented, they may have been among the most taxing possible tasks to be performed by the existing human populations (perhaps 10,000 to 20,000 years ago). It is possible that a rapid burst of "weeding" emerged—i.e., that those would could learn this new, non-innate skill would be differentially able to mate and pass on their genes, and that those who could not learn it were differentially removed from the gene pool. This new grafting of long visual pathways onto the ends of already-long auditory pathways may represent the most recent selectional modification made to human brains—and, given its timing, possibly a modification that somehow made a difference in our competition for resources with other hominids.

And one last factor that is sometimes neglected: the size of human populations. It is noteworthy to consider just how few individuals were involved in these pivotal times for human evolution. Recent estimates place the population of Europe 30,000 years ago at about *5,000 people.* Such low numbers across such a huge area imply that social networks must have been small, and interactions among bands of individuals infrequent. The kinds of community interactions that drive much complex cognitive activity simply didn't exist. Thus there may have been little or no pressure to forge or make use of many of the potential brain pathways among distant association regions of cortex. As the population began to climb, possibly some threshold was crossed beyond which social interactions became dense enough to ignite social cognition, possibly engaging for the first time some of the big brain's latent capabilities.

We can coalesce a number of circumstances, then, that may have doomed the Boskops.

Perhaps, as just argued, they tended toward a slight variant of brain paths, as do some humans, with longer lines in their brains dedicated to visual pathways, but less well-connected for the paths we humans seem to have: long auditory paths in the brain for spoken language, connected with long visual paths to enable writing. As we've seen, individual brain path differences even among humans may underlie different facility with non-innate tasks like reading. Boskops may have been something like extreme examples of some of our individual human differences.

Perhaps instead they excelled at language, and had the tools for writing, but simply didn't thrive long enough to build cultures that would invent and use written language. It apparently took humans tens of thousands of years of language before the invention of writing took hold; other skills were more crucial in the earliest days, when there were no guns, and survival often still went to the strong, wily, and nasty.

We've also shown that large heads are a severe problem in childbirth; humans have far and away the highest rate of death in childbirth, and Boskops' attrition rate would have been far worse, with their slim bodies and huge heads. That may have contributed to their inability to grow large populations. Though they may have thrived in smaller groups, we have argued that these may not have laid sufficient groundwork or sufficient time for extensive cultural expansion, nor its attendant growth of language use.

And even today, in a time when reason can trump violence, we still sometimes lose respected, peace-loving leaders to thugs. Even if the Boskops were the smartest of the hominids, even if they had language, even if they lived among us and were respected for their wisdom, they may have been too fragile for their time. Perhaps they might have been leaders if they were around today, but they may not have survived their 10,000-year-old world of pre-cultural violence.

CHAPTER 14

MORE THAN HUMAN

BRAIN AND SUPERBRAIN

If differently organized brain paths lead to differential faculties, different ways of thinking; and if Boskops had longer, or more integrated pathways, what new capabilities might these longer mental assembly lines have conferred on them?

Long paths: "Think it through"

We have shown in chapter 9 and chapter 13 that longer brain pathways, longer assembly lines, lead to larger and deeper memory hierarchies. These confer greater abilities to examine and discard more blind alleys, to see more consequences of a plan before enacting it. In general these enable us to think things through. If Boskops had longer chains of cortical networks, longer assembly lines, they would have created longer and more complex classification chains. When they looked down a road as far as they could, before choosing a path, they saw farther than we can; more potential outcomes, more possible downstream costs and benefits.

Branching paths: "The continuity of self"

Just as long, straight assembly lines confer certain advantages, so do broad, "bushy" brain paths that may not be as long, but contain large numbers of branches. With multiple paths from a given starting point, overgeneralization is less likely, and individual episodes are more likely to be distinctly stored without being over-categorized. Thus old memories tend to stay intact much longer. These are the things that go into forming our conception of ourselves; if they become obscured by overgeneralization, they slowly obliterate the younger version of one's self. In this scenario, fifty-year-olds no longer feel emotional continuity with their collegiate selves because that rambunctious character, once present in their memories, barely exists there any longer. If Boskops had broad, highly branching brain paths, they may have maintained primary records of their earlier experiences, and hence the ability to recall and re-assemble their younger selves. They may have embodied Faulkner's observation that "The past is never dead. It's not even past."

Intersecting paths: "It's a little like this"

Separate assembly lines (e.g., one for distinguishing facial appearance; one for distinguishing tone of voice; one for analysis of facial expression) that intersect multiple times with each other, will tend to improve the ability to compare and analogize. We describe many new ideas in terms of old ones; analogical reference enables us to compactly convey characteristics of something that is otherwise alien. (Writing a computer program is like building a bicycle.) Broader assembly lines with more branching alternatives, intersecting often with other lines, generate broader and richer memory hierarchies. These generate more analogical matches to events, identifying more ways in which a new memory is like prior ones, even if only tangentially. If Boskops had broader and more intersecting brain connection paths, conversation with them might be a little flowery, possibly even a little annoying—or it might be insightful and eye-opening.

Stations at the ends of paths: "Phaedrus' knife"

Ongoing searches for matching memories identify candidates for analogy, but also identify differences among those candidates. Longer assembly lines cause more of everything in memory, matches and mismatches, to be found. At each cortical processing station in an assembly line, categories can be formed and categories can be split. If Boskops had longer or more intersecting assembly lines, or both, they were probably capable of making very fine distinctions that might easily be lost on minds with shorter brain paths. This may have left them unfit for politics, but they might have made great jurists.

In any event, they died, and we lived, and we can't answer the question, Why? Why didn't they out-think the smaller-brained hominids like ourselves, and spread across the planet? Perhaps they didn't want to. As more possible outcomes of a plan become visible ("think it through"), the variance among judgments between individuals will likely lessen. There are far fewer correct paths, intelligent paths, than there are paths. It is sometimes argued that the illusion of free will arises from the fact that we can't adequately judge all possible moves, with the result that our choices are based on imperfect (sometimes impoverished) infor- mation. Perhaps the Boskops were trapped in their ability to see clearly where things would head. Perhaps they were prisoners of those majestic brains.

There is another, again poignant, possible explanation for the disappearance of the big-brained people. Maybe all that thoughtful- ness was of no particular survival value in 10,000 BC. The great genius of civilization is that it allows individuals to store memory and operating rules outside of their brains, in the world that surrounds them. The human brain is a sort of central processing unit operating on multiple memory disks, some stored in the head, some in the culture. Lacking the external hard drive of a literate society, the Boskops were unable to exploit the vast potential locked up in their expanded cortex. They were born just a few millennia too soon.

NEW PATHS, NEW HUMANS

Traversal paths are determined by the anatomical connection paths in the brain; signals can only go certain ways, and must get to their destinations via the existing paths. As we have suggested, the layout of these paths may determine unique predilections and talents of individuals. The genes that influence the creation of these paths may be selected according to the success of the organisms containing various path arrangements.

In our library analogy, paths to some areas might be more popular than others. For instance, if we tracked the paths of large numbers of library visitors, we might find paths to popular fiction, along a far western wall, much more traveled than paths to, say, political science, in an eastern region of the library.

But these paths may change with the environment. In an era of escapism, fiction might be a more common destination, whereas in an election year, political science might be more well-traveled.

Particular brain paths may be a better fit, then, for some environments than others—and if the environment changes, those previously "optimal" paths may then strain to perform new tasks for which they may be poorly suited. Designs are not optimal in and of themselves. They can be optimal *for* some outcome. A design is only optimal with respect to some particular task; indeed, optimality is undefined without reference to a goal or measure.

In any given population, it may be most useful to have individuals and subpopulations with a range of somewhat different brain path arrangements. Different arrangements will likely lead to differential ability in diverse tasks—so if not all tasks are clearly specified in advance, a healthy distribution of different brains will likely outperform a more homogeneous team.

Slight genetically-steered and environmentally-influenced changes to brain path arrangements might have occurred as the brain was growing large, during the time of the big-brained hominids—from about 200,000 years ago to about 10,000 years ago. Slight changes in brain path layouts might be easily achieved by almost indiscernible gene changes, and yet may have disproportionate effects on cognitive abilities, tendencies, and talents.

Changes in brain path arrangement, when they occur, are likely to leave little or no trace in the fossil record. They may be only barely discernible even in behavior—just a predilection here, a tendency there. But they may have been a factor underlying the brain changes that were occurring as various different kinds of hominids arose.

THE FINAL PATH TO HUMANS

What specific connectivity changes might have occurred in the headlong rush to big brains, that resulted in Idaltu, and Fish Hoek, and Skhul, and Neanderthal, and Boskop, and us?

We have seen that the brain path changes that occurred with early mammalian brain growth probably originated in olfactory circuits, and moved toward auditory circuits. Early mammals were likely small, nocturnal creatures, who needed distance senses like olfaction, and for whom vision, in their night habitats, was less useful. Audition filled the bill—they may have become exquisitely sensitive to the sounds around them, and differential growth of capacity for more and better sound recognition may have been selected for. It has often been noted that the brains of the most primitive mammals were likely dominated by olfaction, as are their likely survivors today: echidnas, setifers, hedgehogs. The next step was probably the thalamo-cortical auditory system, and it may well have been the revolution that took the mammals from just surviving their post-dinosaur world, to thoroughly overtaking that world.

But with that success, the mammals broke free of their nocturnal niches—and abruptly had a use for vision. The visual system, which may have been a relative latecomer to mammalian evolution, soon became a centerpiece of it. Predator and prey alike, mammals grow their visual systems as they grow their brains. Primitive mammals tend to have two monocular eyes, one on each side of their heads, but as the brain grows, mammals' eyes crept to the front of their heads, and the two eyes' inputs became integrated into an ever-more-complex system that could judge distance and three-dimensional depth at a single glance.

The visual system appears to continue its incremental modification with brain growth. By the time primates arrived, they were highly visual animals, and had more specialized machinery built into the dedicated visual parts of their thalamo-cortical circuits than perhaps any other animal.

Perhaps this trend ended with the genus *Homo*.

We think of ourselves as enormously visual creatures. We think of things visually, we use visual analogies, we use terms like "perceiving" and "seeing" almost synonymously. But it is notable that, from the time we are infants, we are intensely, pervasively auditory animals. Human infants are raised to hear, and listen to, and discern, and repeat sounds, from the moment we're born. Most of our learning, from infancy on, is based on language—and that is entirely auditory language; spoken language. To be sure, we learn a vast amount of our world by seeing and doing, trial and error. But most of our uniquely human nature comes from our linguistic links to others.

In previous chapters, we specifically forwarded the hypothesis that a succession of different emphases in the brain have arisen, from the initial adoption of olfaction from reptilian precursors, to the rise of auditory thalamo-cortical circuitry, to the dominance of vision as mammals grew and matured—and then a return to the primacy or near-primacy of audition and its unique use for language.

Other path arrangements in the brain might have arisen in other hominids. Take, for instance, the different skull shape of Neanderthals. They do not have quite the large rising forehead housing an enlarged anterior cortex—they instead have an enlarged "occipital bun" slightly protruding from the back of the brain. Possibly this signals a different arrangement of brain paths, which in turn may have affected their skull shape. As we have mentioned, despite the overwhelming tendency of concerted evolution to determine which areas will grow proportionately larger as the brain grows, nonetheless random variation can generate evolutionary offshoots that happen to have adaptive characteristics, enabling them to survive. If Neanderthals had extensive, branching visual pathways, they might have specialized in making fine visual distinctions, and might even have generated paths that ended up differentially connecting to, and thus specializing in, particular visual phenomena. Just as we have areas that become dedicated to color, and motion, and shape, as well as further downstream areas specializing in faces versus places versus tools, perhaps Neanderthals had many such categories—objects that fit in the hand; objects that could be thrown; objects that one could build with, or fight with— and possibly movement-based categories: objects approaching from an oblique angle; objects moving slowly in the periphery of the visual field. Neanderthals might have been natural baseball batters, or outfielders. Perhaps they would have been gifted artists, or filmmakers.

Boskops may have had pathways more like our own, but variations may well still have arisen—and the talents thus conferred may have been tens of thousands of years before their time. A Boskop uniquely able to write poetry, or novels, or music, or, indeed, to hit a baseball, might have been utterly dysfunctional in the context of his or her world—but might be viewed as unusually gifted if he or she were alive today.

If we could genetically engineer people with different long brain paths, what kinds of diverse abilities might arise?

We have seen that added brain regions, judiciously sited, could have given rise to the qualitative leap from simple symbol use in apes to true language, in all its complexity, in humans. The question is thus raised of what additional capabilities, perhaps currently unimagined, might be birthed if further brain regions were added to existing long brain paths, or if entirely new paths were added—either by natural steps of evolution or by the engineering artifice of man.

Variations in the big intra-cortical pathways between individuals may arise primarily from randomness in the process of wiring up something so complex as a big brain. During development, billions of neurons are growing and competing with each other for targets. Genetic control is an immensely complex affair, in which genes control the timing of neuronal migration into the emerging cortex, with resulting highly-variable struggles among growing neurons. Changes as minor as transient hormonal surges can affect developmental trajectories, and the resulting alterations can have consequences that redound through subsequent brain maturation. Individual differences arising from factors like these, acting after the genes have set things in motion, are quite likely beyond the reach of evolution or of DNA engineers.

The imprecision of these controls can perhaps be seen by analogy. Picture the problem of getting a large crowd out of a baseball stadium. The builders who designed the stadium knew that 40,000 people would be using it for any given game, and so included a certain number of exits carefully placed at particular locations. But they did not try to funnel fans in section K to exit #7. Why not? Because the crowd will inevitably vary from day to day, and chance events will influence the outcome (a fan blocks an outfielder's play; a rally is ended; fans start leaving by the seventh inning). A more efficient solution to emptying the park is simply to let the members of the crowd interact with each other and thereby find their own way out. Local interactions and decisions obviate the need for detailed top-down designs, and in fact work better. Indeed, a science of complex systems has developed to study the variation inherent in circumstances like these, and the limitations of preordained designs applied to living systems.

It is notable that some local developmental interactions render the individual unviable, and thus never get born; and some gene controls coerce development in certain directions. This makes certain types of variation much more likely than others. The result, as pointed out, is that not all random variants can occur: there are no mammals with three ears, or mouth above the nose, let alone tentacles or millions of other never-seen variants. Variation is funneled into categories. Thus some potential manipulations will likely be ineffective in building different brains; many variations likely won't "take," given the brain's developmental and genetic constraints.

If profoundly useful and transformative linguistic abilities arose almost full-blown via the brain expansion from ape to human, might there also be leaps of equal size if artificial brain systems, or robot brains, are engineered to the size of human brains, and beyond?

This specter has been explored in the realm of science fiction, but in our understanding of brain paths, and their specific conferral of language abilities, spliced on to pre-human brain systems, perhaps a glimpsed route to future new capabilities is revealed.

INCONSTANT BRAIN

Dogs—the docile, domesticated, trainable, utterly trustworthy friends of humanity—were once wolves—unpredictable, wily, wild, dangerous killers. Debates abound on questions of how long their domestication took. When did humans first tame wolves? How many generations of selective breeding before we arrived at collies and poodles?

Imagine that you could go back in time to the first tentative approaches between wolves and man. What were the first friendly interactions? What were the steps to new gene pools for dogs? What were those first wolf-dogs like? Did they look like dogs? Were there already some more-friendly and some less-friendly wolves? Would the children of friendly wolves tend also to be friendly? Would they have other traits? Are dogs more or less "intelligent" than wolves?

We may never know the answers. But amazingly, the experiment has been run again, quite recently. A Russian scientist in the 1950s named Dmitry K. Belyaev had been interested in genetics and intensive breeding of animals. His career had been all but shut down by the Soviets, who found his ideas subversive. He moved to the wilds of Siberia and embarked on a grand and quixotic journey: he decided to domesticate foxes.

Belyaev founded the Siberian Department of the Soviet (now Russian) Academy of Sciences, collected 130 foxes (100 females, 30 males), and began breeding them selectively—for docility. In any given generation, only the tamest few foxes were allowed to breed. No training was performed; the only manipulation was selective breeding. From 1959 until his death in 1985, he bred more than 45,000 foxes over roughly 30 to 35 generations.

To what effect? Each pup was tested from age one month on, once a month. An experimenter offered food from his or her hand, while trying to stroke or pet the pup. Pups that either recoiled or attacked were assigned to a group Belyeav simply called Class III. Those that allowed themselves to be petted, but otherwise showed no particularly friendly response to humans, were Class II. And foxes in Class I? These pups were overtly friendly, literally whimpering to attract attention and sniffing and licking experimenters, like dogs. They also wagged their tails. They craved human contact. Even those that escaped from the farm eventually returned on their own. In other words, they are as different from foxes as dogs are from wolves.

Inside thirty generations, Belyaev created a new species.

We can ask what is different in these new creatures. Remember: no training has taken place, yet their behavior is markedly different. Clearly there are differences in their brains. Just as clearly, these differences arose from selective changes in their genes. It is helpful to note that there are observable differences in the rest of their bodies as well. They have coats of different colors whereas normal foxes are solidly grey or red. They tend to have floppy ears, unlike foxes. They sometimes have rolled tails, white body and face markings, slightly shorter legs.

Their brains do show chemical differences: higher levels of certain neurotransmitters such as serotonin. But no one has yet looked in detail at these brains: their differential function during behavior, nor differential connectivity.

Genes may indeed have changed, but were they in the brain? Belyaev may have altered body genes that secondarily influenced brain transmitter levels. As noted, the Class I animals look different, with different colors, ears, tails, and face markings, indicating that the breeding modified genetic programs controlling the development of multiple body parts and hormonal systems, effects that would undoubtedly modify maturation of the brain. We may not get a chance to look inside their brains: the experiment in Russia has waning funds, and the researchers have embarked on a sales attempt, selling domesticated foxes to raise money. At this writing, their fate is unsure.

This episode is reminiscent of a longer history of attempts to build different, possibly better, brains through breeding. Selective mating studies in the 1930s produced rats with spectacular scores on maze tests; these were assumed to be due to genetic modification of the brain. But it was later found that simply putting normal rats in a complex environment for a few weeks (a treatment that certainly doesn't modify DNA codes) produced similar effects. Nutrition, hormones, blood circulation, and myriad other variables influence the activity of brain genes, and all are sensitive to genetic inheritance and to environmental variables. As noted, through all this work, there has yet to be evidence for changes in the relative proportions of brain parts (see chapter 11).

That experiment bred for docility. But who knows what other traits might be bred for? And who knows what animals might be bred?

NEXT STEPS

It could start today. Find a group of individuals who have a set of distinguishing genetic profiles in common. Possibly they have a high-functioning version of Asperger's syndrome; and perhaps their brain paths confer an unusual ability to write large computer programs. (A link of just this kind has been proposed by scientists who study autism and Asperger's.) They marry and have kids. Perhaps they decide to move to an isolated island—one with internet access—so that they can continue to program, but have no other interaction with other members of society.

They then fit the profile of a potential new species. They share a trait that is likely to set them apart from many other humans, they interbreed, and they isolate themselves. As we discussed in chapters 3 and 10, no one knows exactly how a subgroup "speciates"—i.e., becomes its own separate species, losing its ability to interbreed with members of its previous species. After generations, these individuals might become a separate species—separate from the rest of humanity, beginning a new evolutionary path, generating their own independent genetic variations.

Not likely. But there are other ways.

Genetic variants might be artificially engineered. It is possible—though very likely not desirable—to modify the genes of a group of people, via the same principles as "gene therapy," in which doctors hope to change the genes of those with predilections for certain diseases. It may even become possible to modify the genes of a subpopulation such that they would no longer be able to mate with other humans, but could mate only with each other. Genetic variants could potentially be engineered with different brain path arrangements, uniquely suited to particular abilities, like programming computers, or composing music, or possessing unusually large memory capacity.

There are groups that might actively advocate such a plan—"extropians" and "transhumanists," who already promote changes to enable humans to surpass typical limits of longevity and capability.

Possibly, such methods wouldn't be used to create a single "superman," but rather, to create different brain paths of many different kinds: individuals uniquely suited for individual pursuits. If we could promote the brain growth of different connectivity paths—longer versions of ones we have now, new ones not seen in humans, combinations of these—we might add depth and complexity to our existing abilities, and might arrive at novel abilities, new specializations, new insights, that weren't suspected.

Asperger's mutations are likely acting on cortical developmental events, with yoked consequences far beyond enhanced computer programming skills. In general, attempts to exploit known genetic variations in current human populations may result in too may unforeseen ancillary variations. But gene methods are also being applied to the individual machinery used by synapses to encode memory. These experiments are still limited to laboratory animals, but the results have been both selective, without side effects, and

effective: by changing single proteins pivotal to learning, the resulting animals exhibit normal everyday behavior but are exceedingly
bright. Science may well identify genes that can be modified to produce an animal, even a human, that will learn faster without unacceptable side effects. This is the future that likely awaits us. It is a
prospect that dwarfs daily concerns, and it would be a good idea for
humanity to start thinking about it now. In fact, these prospects are
with us now, in an early form; not from gene changes but from chemical manipulation. Gene research on synapses is itself based on the
remarkable successes of brain science in unraveling the immense
complexity of the synaptic learning machine; those same successes
have fueled the pharmaceutical industry's search for drugs that modify the same proteins targeted by DNA workers. Drugs have been
built that selectively enhance the brain's ability to collect, assemble,
and encode information—and these effects have been reported in
many laboratories, in rats, in monkeys, and in humans.

Two comparisons with big-brained Boskops are immediately
apparent. The new drugs improve communication between cortical
neurons, which may allow brain regions to assemble larger than
normal networks, and to find paths to other areas, thus improving
essential steps in building the assembly lines underlying thought.
Boskops likely had these communication advantages due to their
added neurons and taller dendritic trees; the drugs may achieve
some of the effect via a different route. And compounds have been
invented that turn on brain growth factor genes, which won't
change brain size but may well slow the neuronal wear and tear that
comes with aging. Boskops' larger brains likely had higher memory
capacity; enhanced growth factor activity may enable humans to
retain a greater percentage of their original memory capacity well
past middle age.

Perhaps the Boskops, or their like, could be recreated. These
Neo-Boskops might be smarter than us in many ways, and yet, as
just described, they could be far more diverse than we are, exhibiting
extremes of differential faculties.

During their time on earth, the Boskops may well have been more
peaceable than we were, and we likely exterminated them. But that

may be less likely this time around. It used to be possible for outsiders to invade an indigenous population, and to eliminate them,
even if the invaders were no smarter than the invaded—as may have
been the case when Alaric the Visigoth arrived at the gates of Rome
in 410 AD, or when Europeans arrived en masse in the Americas in
the fifteenth century. Without adequate defense technologies, overwhelming numbers could outweigh other superiorities. But in a
current or future world of relatively level playing fields, perhaps
Neo-Boskops could arrive, and be tolerated as newcomers to our
society.

In general, those who believe in exclusivity or totalitarianism,
fanaticism, or terrorism, won't want such visitors. A population that
invites inclusivity, intellect, and diversity, could welcome them.
Perhaps they'll outshine us as orchestra conductors, or baseball
pitchers, or ornithologists. Perhaps they'll come and this time
survive. Perhaps they'll have things to teach us. Perhaps we'll learn.

CODA

In any event, Boskops are gone, and the more we learn about them,
the more we miss them. Their demise was likely to have been
gradual. A big skull was not conducive to easy births, and thus a
within-group pressure toward smaller heads was probably always
present, as it is still is in present-day humans, who have an unusually
high infant mortality rate due to big-headed babies. This pressure,
together with possible interbreeding with migrating groups of
smaller-brained peoples, may have led to a gradual decrease in the
frequency of the Boskop genes in the growing population of South
Africa.

Then again, as is all too evident, human history has often been
a history of savagery. Genocide and oppression seem primitive,
whereas modern institutions from schools to hospices seem
enlightened. Surely, we like to think, our future portends more of
the latter than the former. If learning and gentility are signs of
civilization, perhaps our almost-big brains are straining against

their residual atavism, struggling to expand. As we've suggested, perhaps the preternaturally-civilized Boskops had no chance against our barbarous ancestors, but could be leaders of society if they were among us today.

Maybe traces of Boskops, and their unusual nature, linger on in isolated corners of the world. Physical anthropologists report that Boskop features still occasionally pop up in living populations of Bushmen, raising the possibility that the last of the race may have walked the dusty Transvaal in the not-too-distant past. As we pointed out in chapter 10, some genes stay around in a population, or mix themselves into surrounding populations via interbreeding. The genes may remain on the periphery without becoming widely fixed in the population at large, nor being entirely eliminated from the gene pool.

Just about 100 miles from the original Boskop discovery site, further excavations were once carried out by none other than our indefatigable museum director, Frederick FitzSimons. He knew what he had discovered, and was eagerly seeking more of these skulls. At his new dig site, he came across a remarkable piece of construction. The site had been at one time a communal living center, perhaps tens of thousands of years ago. There were many collected rocks, leftover bones, and some casually-interred skeletons of normal-looking humans. But to one side of the site, in a clearing, was a single, carefully-constructed tomb, built for a single occupant; perhaps the tomb of a leader, or of a revered wise man. His remains had been positioned to face the rising sun. In repose, he appeared unremarkable in every regard . . . except for a giant skull.

APPENDIX

Big brains are a big topic; the notes and references could readily fill hundreds of pages. We instead list primary readings that either were the earliest to appear on particular discoveries, are highly cited in the field, or provide exceptional reviews. The result is a highly selective scholarly bibliography, intended to suggest fruitful starting points for further reading.

CHAPTER 1

A brain is typically measured either in cubic centimeters of volume (cc), or in grams (g) of weight; these measurements are identical for water, and brains have roughly the same density as water.

The following table lists some of the most well-studied human-like skulls, all with capacities larger than the human norm of 1350 cc.

Name	Location	~size (cc)	~age (Kya)
Idaltu	Ethiopia	1450	150
Omo II	Ethiopia	1435	130
Singa	Singa, Sudan	1550	130
Skhul V	Mt Carmel, Israel	1520	115
Skhul IV, IX	Mt Carmel, Israel	1590	115
Qafzeh 6	Nazareth, Israel	1565	95
Qafzeh 9	Nazareth, Israel	1508	90
Border Cave	Natal, S.Africa	1510	75
Cro-Magnon 1	Dordogne, Fr	1600+	30
Brno 1	Brno, Czech	1600	25
Fish Hoek	Skildergat, S.Africa	1600	12

Zhoukoudian	China	1500	10
Wadjak	Java, Indonesia	1550	10
Tuinplaas	Pretoria, S.Africa	1590	?
Boskop	Transvaal, S.Africa	1717+	?

The list is ordered by estimated age of the skull (right-hand column). It can be seen that the skulls are already suprahuman in size by about 150 thousand years ago with *Idaltu*, and continue to grow straight through near-modern times, arriving at the huge South African skulls of Fish Hoek, Tuinplaas, and Boskop within the last fifteen thousand years. All of these skulls are strikingly human-like, with broad, rising foreheads, and relatively small jaws and faces.

(For reference, two of the most prominent large skulls of the Neanderthal type are listed here):

| Shanidar | Zagros, Iraq | 1600 | 60 |
| Amud | Wadi Amud, Israel | 1740 | 45 |

An admirably comprehensive and attractive compendium of human and hominid skulls is found in:

Schwartz, J., Tattersall, I. (2003). The Human Fossil Record, Vols 1–4. Wiley.

One of the oddities of the Boskop story is the disconnect between the rich trove of references from the early twentieth century, and the paucity of references after that time. Some of the best accounts can be found in:

FitzSimons FW (1915). Palaeolithic man in South Africa. *Nature*, 95: 615–616.

Broom R (1918). The Evidence Afforded by the Boskop Skull of a New Species of Primitive Man (*Homo Capensis*). *Anthropological Papers of the American Museum of Natural History*, 23: 65–79.

Galloway A (1937). The Characteristics of the Skull of the Boskop Physical Type. *American Journal of Physical Anthropology*, 32: 31–47.

Pycraft W (1925). On the Calvaria Found at Boskop, Transvaal, in 1913, and Its Relationship to Cromagnard and Negroid Skulls. *Journal of the Royal Anthropological Institute of Great Britain and Ireland*, 55: 179–198.

Tobias P (1985). History of Physical Anthropology in Southern Africa. *Yearbook of Physical Anthropology*, 28: 1–52.

Haughton S (1917). Preliminary note on the ancient human skull remains from the Transvaal. *Transactions of the Royal Society of South Africa*, 6: 1–14.

Dart R (1923). Boskop remains from the south-east African coast. *Nature*, 112: 623–625.

Dart R (1940). Recent discoveries bearing on human history in southern Africa. *Journal of the Royal Anthropological Institute of Great Britain and Ireland*, 70: 13–27.

Many other animals make a few tools and buildings; notably, other primates such as chimps, certain birds such as crows, some other mammals such as beavers. None makes anything like our array of specialized tools, differentiated dwellings, and other inventions. Other animals such as chimps arguably pass on cultural information, handing down methods from generation to generation; again, these intriguing and exciting findings can nonetheless readily be seen to be enormously different in scope from the ability of humans to pass information via language.

The notable size of cetacean brains may be reason to expect comparably exceptional intelligence. This is a field of inquiry in which further discoveries may be forthcoming.

CHAPTER 2

The original Dartmouth Conference on Artificial Intelligence was held in 1956 at Dartmouth College in Hanover, NH. A fifty-year retrospective of the meeting was recently held at Dartmouth; details can be viewed at:

http://www.dartmouth.edu/~ai50

Computational science has as its goal to understand phenomena sufficiently well to reconstruct them. This principle underlies McCarthy's statement, and the same principle was adhered to by the well-known physicist Richard Feynman; on his blackboard at the time of his death in 1988 was this statement: "What I cannot create, I do not understand." (in: Hawking S (2001). The Universe in a Nutshell. Bantam. p.83.)

Initial publications describing the computational analysis of brain circuits focused on the olfactory system:

Ambros-Ingerson J, Granger R, Lynch G. (1990). Simulation of paleocortex performs hierarchical clustering. *Science*, 247: 1344–1348.

Granger R, Staubli U, Powers H, Otto T, Ambros-Ingerson J, Lynch G. (1991). Behavioral tests of a prediction from a cortical network simulation. *Psychological Science*, 2: 116–118.

McCollum J, Larson J, Otto T, Schottler F, Granger R, Lynch G. (1991). Short-latency single-unit processing in olfactory cortex. *Journal of Cognitive Neuroscience*, 3: 293–299.

Karel Capek's play first appeared on stage in Prague in 1921:

Capek K (1920) RUR: Rossum's Universal Robots (Rossumovi Univerzalni Roboti).

Recent publications describing new "non-von" computer designs based on brain architectures:

Furlong J, Felch A, Nageswaran J, Dutt N, Nicolau A, Veidenbaum A, Chandreshekar A, Granger R. (2007). Novel brain-derived algorithms scale linearly with number of processing elements. Proceedings of the International Conference on Parallel Computing, (parco.org) 2007.

Moorkanikara J, Chandrashekar A, Felch A, Furlong J, Dutt N, Nicolau A, Veidenbaum A, Granger R (2007). Accelerating brain circuit simulations of object recognition with a Sony PlayStation 3. International Workshop on Innovative Architectures (IWIA). More information can be found at brainengineering.org

CHAPTER 3

A vivid introduction to genes in evolution and in development by Sean Carroll can be found in:

Carroll S. (2005). Endless forms most beautiful. The new science of evo devo and the making of the animal kingdom. NY: WW Norton.

An excellent introduction to the genetic underpinnings of the brain can be found in:

Marcus G. (2004). The birth of the mind: How a tiny number of genes creates the complexities of human thought. NY: Basic Books.

Niles Eldredge and Stephen Jay Gould's introduction of punctuated equilibria:

Eldredge N, Gould S. (1972) Punctuated equilibria: an alternative to phyletic gradualism. In: *Models In Paleobiology* (Ed. by T. J. M. Schopf). Freeman Cooper.

Stephen Jay Gould's book on the Burgess Shale:

Gould SJ (1990). Wonderful life: The Burgess Shale and the nature of history. NY: WW Norton.

Examples have been presented that some human (and other animal) features can be viewed as being optimized with respect to certain tasks (e.g., Changizi M (2003). The brain from 25,000 feet. Springer.) As mentioned, optimization is not universal; a mechanism is only optimized *with respect to* some particular outcome. Our body parts perform multiple very-different tasks (e.g. our hands grasp, climb, hit, groom, throw; we have organs that share tasks of copulation and urination; we have separate circulatory systems for blood and lymph fluid), and thus may be compromises capable of participating in many tasks but less than optimal at any particular one.

Very small parts of the brain (restricted regions of a phylogenetically ancient component of the hippocampus) continue to grow new neurons into adulthood.

> Gage F (2002) Neurogenesis in the adult brain. *Journal of Neuroscience* 22: 612–613.

There have been controversial reports of neocortical neuron growth (though even these apparent new neurons died within a few weeks after being generated):

> Gould E, Reeves A, Graziano S, Gross C. (1999). Neurogenesis in the Neocortex of Adult Primates. *Science*, 286: 548–552.

Subsequent publications raised significant doubts about the validity of these findings:

> Kornack D, Rakic P (2000). Cell proliferation without neurogenesis in adult primate neocortex. *Science*, 294: 2127–2130.

And, more recently, a set of extremely clever and sensitive experiments have shown that the adult human neocortex simply *does not contain any neurons acquired after early development.*

> Bhardwaj R, Curtis M, Spalding K, Buchholz B, Fink D, Bjork-Eriksson T, Nordborg C, Gage F, Druid H, Eriksson P, Frise J (2006). Neocortical neurogenesis in humans is restricted to development. *Proceedings of the National Academy of Sciences*, 103: 12564–12568.

These remarkable experiments are worth recounting. How would you know for sure whether new neurons existed in the brain? If we could invent a fanciful experiment, we might imagine injecting a long-lasting radioactive (slowly decaying) measurable label into the neurons of someone's brain when they're born, then waiting for them to grow through adulthood, and then, after they die, to check millions of their neocortical neurons to see whether any of them do not contain the marker, which would provide evidence that those neurons were younger than the brain they were in, and thus that those neurons had been generated after the person's childhood. Bhardwaj et al. carried out just this experiment. A radioactive isotope of carbon, C14, is created by nuclear bombs; many of these were tested in the 1940s and 1950s; the levels of C14 in the atmosphere, in the food chain, and ultimately in people, are very accurately measurable, and changed markedly from year to year. Thus the "birthday" of any cell can be accurately determined by measuring the amount of C14 in that cell. The above scientific team led by Ratan Bhardwaj at the Karolinska Institute in Stockholm, Sweden, and including collaborators from many other institutions, measured millions of neocortical neurons in the brains of seven individuals of known ages. Put starkly, they showed that none of the neurons in their brains were younger than the individuals; that is, no new neurons had been born in the neocortex after these individuals' childhoods.

On variation among human genomes:

Redon R, Ishikawa S, Fitch K, Feuk L, and 39 additional authors (2006). Global variation in copy number in the human genome. *Nature*, 444: 444–454.

CHAPTER 4

Many recommended texts for the basics of neuroscience:

Kandel E, Schwartz J, Jessell T (2000). Principles of neural science, 4th ed., NY: McGraw-Hill.

Swanson L (2002). Brain Architecture: Understanding the basic plan. Oxford University Press.

Jerison H. (1974) Evolution of the Brain and Intelligence. Academic Press.

Shelley M (1818). Frankenstein, Or, the modern Prometheus. London: Lackington Hughes.

On the regularity of neocortex:

Rockel AJ, Hiorns RW, Powell TP (1980) The basic uniformity in structure of the neocortex. Brain 103:221–244.

Swanson L (2002). Brain Architecture: Understanding the basic plan. Oxford University Press.

CHAPTER 5

A comprehensive and highly readable scientific discussion of brain evolution is found in:

Striedter G (2005). Principles of Brain Evolution. Sinauer Associates.

Discussion of the brain's low-precision components achieving high-precision performance can be found in:

Granger R (2005). Brain circuit implementation: High-precision computation from low-precision components. In: Replacement Parts for the Brain (T.Berger, D.Glanzman, Eds.) MIT Press, pp. 277–294.

Granger R (2006). Engines of the brain: The computational instruction set of human cognition. AI Magazine 27: 15–32.

Much has been written on synaptic change and long-term potentiation (LTP). Discussion and review can be found in:

Baudry M, Davis J, Thompson R (1999). Advances in synaptic plasticity. MIT Press.

Bliss T, Collingridge G (1993). A synaptic model of memory: Long-term potentiation in the hippocampus. Nature, 361: 31–39.

CHAPTER 6

On striatal implants, controlling bulls in bullfights, and controlling emotions via amygdala implants:

Delgado J (1969) Physical control of the mind: Toward a pscyhocivilized society. Harper & Row.

Crichton M. (1972). The Terminal Man. A.Knopf.

Original report of the epileptic patient "Henry," and his memory impairment after brain surgery:

Scoville W, Milner B (1957). Loss of recent memory after bilateral hippocampal lesions. Journal of Neurology, Neurosurgery, and Psychiatry, 20: 11–21.

William James's description of a child's impression of the world as "a great blooming, buzzing confusion" is from:

James W (1890). The Principles of Psychology. Boston: Henry Holt. (p. 462)

The recent renaming of almost all primary structures in the avian brain:

Reiner et al. (2004) Revised nomenclature for avian telencephalon and some related brainstem nuclei. J. Comp Neurol., 473: 377–414.

Seminal integrative work on the brain:

Szentagothai J (1975) The 'module-concept' in cerebral cortex architecture. Brain Research 95:475–496.

Valverde F (2002) Structure of the cerebral cortex. Intrinsic organization and comparative analysis of the neocortex. *Revista de Neurologia*, 34:758–780.

Sherrington C (1906). The integrative action of the nervous system. NY: Scribner.

Olfaction and cortical evolution:

Lynch G (1986). Synapses, circuits and the beginnings of memory. MIT Press.

Aboitiz F, Morales D, Montiela J (2003). The evolutionary origin of the mammalian isocortex: Towards an integrated developmental and functional approach. Behavioral and Brain Sciences, 26: 535–586.

CHAPTER 7

Much has been written on the residual ur-mammal brain underlying the human forebrain, its effects on our behavior, and the effects on it of many centrally-active drugs.

The great English neurologist John Hughlings-Jackson observed, in the waning days of the 19th century, that epilepsy was preceded by the sudden appearance of odd movements and behaviors. He concluded that the epilepsy "silenced" of the cortex, "releasing" lower brain regions that the cortex had kept in check. As brain size expands and the cortex becomes disproportionately larger, more functions appear to be "taken up" by the cortex that are carried out subcortically in smaller brains. It is difficult to detect motor problems in a rat even when its motor cortex is damaged, but the same injury in a human leaves the subject paralyzed (this often occurs in strokes). Similarly, a rat without a visual cortex still acts visually capable, whereas a human without a visual cortex is irreversibly blind. As these examples illustrate, the owner of a massive cortex becomes completely dependent upon it. See:

Jackson JH (1925) Neurological fragments. London: Oxford University Press.

Critchley M, Critchley E (1999). John Hughlings Jackson. Oxford University Press.

The movie Forbidden Planet (1956) starred Walter Pidgeon, Anne Francis, and a young Leslie Nielsen.

CHAPTER 8

The effect of our learned expectations on perception can give rise to some kinds of visual illusions, such as those in "change blindness," in which we literally fail to see changes occurring to a picture right before our eyes, due to our preconceptions.

Levin D, Simons D (1997) Failure to detect changes to attended objects in motion pictures, *Psychonomic Bulletin and Review* 4: 501–506.

A good treatise on the re-creation of images in the brain:

Kosslyn S, Thompson W, Ganis G (2006). The case for mental imagery. Oxford University Press.

On hidden steps in apparently-simple recognition, and successive recognition of categories and individuals:

Mervis C, Rosch, E. (1981) Categorization of natural objects. *Annual Review of Psychology* 32:293–299.

Rodriguez A, Whitson J, Granger R. (2004), Derivation and analysis of basic computational operations of thalamocortical circuits. *Journal of Cognitive Neuroscience*, 16:856–877.

Liu J, Harris A, Kanwisher N (2002) Stages of processing in face perception: an MEG study. *Nature neuroscience*, 5: 910–916.

Grill-Spector K, Kanwisher N (2005). Visual recognition: As soon as you know it is there, you know what it is. *Psychological Science*, 16: 152–160.

On sequences, hierarchies of sequences of categories, and brain grammars:

Granger R, Whitson J, Larson J, Lynch G (1994) Non-Hebbian properties of long-term potentiation enable high-capacity encoding of temporal sequences. *Proceedings of the National Academy of Sciences*, 91: 10104–10108.

Granger R. (2006) Engines of the brain: The computational instruction set of human cognition. *AI Magazine* 27: 15–32.

Ramus F, Hauser M, Miller C, Morris D, Mehler J (2000). Language discrimination by human newborns and by cotton-top tamarin monkeys. Science, 288: 349–351.

Much has been written recently on grandmother cells versus distributed representations:

Haxby J (2006). Fine structure in representations of faces and objects. *Nature Neuroscience* 9: 1084–1086.

Grill-Spector K, Sayres R, Ress D (2006). High-resolution imaging reveals highly selective nonface clusters in the fusiform face area, *Nature Neuroscience* 9: 1177—1185.

Reddy L, Kanwisher N (2006). Coding of visual objects in the ventral stream. *Current Opinion in Neurobiology* 16: 408–414.

On the "enchanted loom:"

Sherrington C (1906). The integrative action of the nervous system. NY: Scribner.

Changes to specialized regions during development:

Golarai1 G, Ghahremani1 D, Whitfield-Gabrieli S, Reiss A, Eberhardt J, Gabrieli J, Grill-Spector K (2007). Differential development of high-level visual cortex correlates with category-specific recognition memory. *Nature Neuroscience* 10: 512–522.

CHAPTER 9

Description of language abilities in Williams syndrome:

Bellugi U, Wang P, Jernigan T (1994). Williams syndrome: An unusual neuropsychological profile. In: S.Broman, J.Grafman, eds., Atypical cognitive deficits in developmental disorders. NJ: Erlbaum.

Karmiloff-Smith A (1998). Development itself is the key to understanding developmental disorders. *Trends in Cognitive Sciences*, 2: 389–398.

Different brain path wiring in people with higher and lower reading skills:

Klingberg T, Hedehus M, Temple E, Salz T, Gabrieli J, Moseley M, Poldrack R (2000). Microstructure of temporo-parietal white matter as a basis for reading ability: Evidence from diffusion tensor magnetic resonance imaging. *Neuron*, 25: 493–500.

Deutsch GK, Dougherty RF, Bammer R, Siok WT, Gabrieli JD, Wandell B (2005). Children's reading performance is correlated with white matter structure measured by diffusion tensor imaging. *Cortex*, 41: 354–363.

Ben-Shachar M, Dougherty R, Wandell B. (2007). White matter pathways in reading. *Current Opinion in Neurobiology, 17: 1–13.*

CHAPTER 10

The naming schemes for apes and humans continue to be in flux. Use of the term Hominidae is now often supplanted by new terms in which Hominini include the genus *Homo* and Pan (chimps); Homininae includes those plus gorillas; Hominidae includes monkeys (genus Pongo) and Homonoidea refers to all these plus gibbons (Hylobates). A sub-tribe of "hominians" denotes only humans and our now-extinct nearest relatives. The term hominid is still often used to refer to just this subgroup, as we do here.

There are many additional fossils that are candidate members of the genus *Homo*, such as the small "hobbit" fossils, currently termed *Homo floresiensis*. Their relationship with humans is still unknown. The familiar "cro-magnon" is typically cited as an instance of a thoroughly modern *Homo sapiens*. Modern indeed; as we mentioned in the notes to chapter 1, the skulls are roughly 30,000 years old and yet have cranial capacities far larger than those of most modern humans.

Caramelli D, Lalueza-Fox C, Vernesi C, Lari M, Casoli A, Mallegni F, Chiarelli B, Dupanloup I, Bertranpetit J, Barbujani G, Bertorelle G (2003) Evidence for a genetic discontinuity between Neandertals and 24,000-year-old anatomically modern Europeans. *Proceedings of the National Academy of Sciences*, 100: 6593–6597.

Seminal articles on the surprisingly recent genealogical relationships between chimps and humans:

Hobolth A, Christensen O, Mailund T, Schierup M (2007) Genomic Relationships and Speciation Times of Human, Chimpanzee, and Gorilla Inferred from a Coalescent Hidden Markov Model. *PLoS Genetics 3*(2): e7 17319744 doi:10.1371/journal.pgen.0030007

Patterson N, Richter DJ, Gnerre S, Lander ES, Reich D (2006) Genetic evidence for complex speciation of humans and chimpanzees. *Nature* 441: 1103–1108.

Enard W, Paabo S. (2004). Comparative primate genomics. *Annual Reviews: Genomics and Human Genetics*, 5: 351–378.

A useful reference on concepts of race:

Brace CL (2005). "Race" is a four-letter word: The genesis of the concept. Oxford University Press.

Schwartz J (2006) Race and the odd history of human paleontology. The Anatomical Record, 289B: 225–240.

CHAPTER 11

An engaging version of the story of Ernst Haeckel and Eugene DuBois is found in Pat Shipman's book:

Shipman P (2002) The man who found the missing link: Eugene Dubois and his lifelong quest to prove Darwin right. Harvard University Press.

Finlay and Darlington's article detailing the "late equals large" hypothesis:

Finlay B, Darlington R (1995). Linked regularities in the development and evolution of mammalian brains. *Science*, 268: 1578–1584.

Data on brain size:

Stephan H, Bauchot R, Andy OJ (1970) Data on size of the brain and of various brain parts in insectivores and primates. In: Noback C, Montagna W (Eds) Advances in primatology, V.1. Appleton. pp. 289–297.

Stephan H (1972). Evolution of primate brains: a comparative anatomical approach. In: R.Tuttle (Ed), Functional and Evolutionary Biology of Primates, Aldine-Atherton. pp. 155–174.

Stephen H, Frahm H, Baron G (1981). New and revised data on volumes of brain structures in insectivores and primates. *Folia Primatologica*, 35: 1–29.

McHenry H (1992). Body size and proportions in early hominids. American Journal of Physical Anthropology, 87: 407–431.

McHenry H (1994). Tempo and mode in human evolution. *Proceedings of the National Academy of Sciences*, 91: 6780–6786.

The idea that brains grew abruptly larger about two million years ago, in concert with learning about tools, was fancifully captured in the opening scenes of 2001: A Space Odyssey, by Stanley Kubrick and Arthur C. Clarke. A *Homo habilis* with a newly enlarged brain receives the insight that a tool can also be a weapon. It clearly was a bad day for *Australopithecus*.

An intriguing publication describing relationships between upright posture and childbirth:

Trevathan, W. (1996). The Evolution of Bipedalism and Assisted Birth. *Medical Anthropology Quarterly*, 10: 287–290.

Baby size and brain size:

Lynch G, Hechtel S, Jacobs D (1983). Neonate size and evolution of brain size in the anthropoid primates. J. Human Evolution, 12: 519–522.

It is again worth noting the large size of cetacean brains. With no constraints on their hip breadth, the birth canal poses no impediment to large-headed babies. Another accident, and again a reason to think that perhaps cetaceans might indeed be expected to be unusually intelligent.

Marino L, Uhen M, Pyenson N, Frohlich B (2003). Reconstructing cetacean brain evolution using computed tomography. *The Anatomical Record*. 272B: 107–117.

Standing's experiments on memory capacity:

Standing L (1973). Learning 10,000 pictures. *Quarterly Journal of Experimental Psychology*. 25: 207–222.

CHAPTER 12

Calculation of expected brain sizes uses a typical brain-to-body equation for the hominoidae:

$$y = 0.74x + 2.2$$

Howard Gardner on multiple intelligences:

Gardner H. (1993). Frames of Mind: The theory of multiple intelligences. Basic Books.

The naturalist Loren Eiseley on the Boskop skull:

Eiseley L. (1958) The Immense Journey. London: V.Gollancz.

CHAPTER 13

We refer again to references from Striedter (from chapter 5) and Finlay (chapter 11).

On microcephalin and ASPM:

Dorus S, Vallender EJ, Evans PD, Anderson JR, Gilbert SL, Mahowald M, Wyckoff GJ, Malcom CM, Lahn BT. (2004). Accelerated evolution of nervous system genes in the origin of *Homo sapiens*. *Cell*, 119:1027.

Evans P, Gilbert S, Mekel-Bobrov N, Vallender E, Anderson J, Vaez-Azizi L, Tishkoff S, Hudson R, Lahn B. (2005). Microcephalin, a gene regulating brain size, continues to evolve adaptively in humans. Science, 309: 1717–1720.

Mekel-Bobrov N, Gilbert S, Evans PD, Vallender E, Anderson J, Hudson R, Tishkoff S, Lahn B. (2005) Ongoing adaptive evolution of *ASPM*, a brain size determinant in *Homo sapiens*. *Science*, 309: 1720.

On brain shape and allometry:

Semendeferi K, Lu A, Schenker N, Damasio H (2002). Humans and great apes share a large frontal cortex. *Nature Neuroscience*, 5: 272–276.

Holloway R (2002). Brief communication: How much larger is the relative volume of area 10 of the prefrontal cortex in humans? *American Journal of Physical Anthropology*, 118: 399–401.

On regional anatomical differences:

Buxhoeveden D, Switala AE, Litaker M, Roy E, Casanova M. (2001) Lateralization in human planum temporale is absent in nonhuman primates. *Brain Behavior and Evolution*, 57: 349—358.

Buxhoeveden D, Switala A, Roy E, Litaker M, Casanova M. (2001) Morphological differences between minicolumns in human and non human primate cortex. *American Journal of Physical Anthropology*, 115: 361–371.

Sherwood C, Broadfield D, Holloway R, Gannon P, Hof P. (2003). Variability of Broca's area homologue in African great apes: Implications for language evolution. *The Anatomical Record*. 271A: 276–285.

Preuss T, Qi H, Kaas J. (1999) Distinctive compartmental organization of human primary visual cortex. *Proceedings of the National Academy of Sciences*, 96: 11601–11606.

On the FOXP2 gene and potential implications for language:

Lai C, Fisher S, Hurst J, Levy E, Hodgson S, Fox M, Jeremiah S, Povey S, Jamison D, Green E, Vargha-Khadem F, Monaco A (2000). The SPCH1 region on human 7q31: Genomic characterization of the critical interval and localization of translocations associated with speech and language disorder. *American Journal of Human Genetics*, 67: 357–368.

Vargha-Khadem F, Gadian D, Copp A, Mishkin M (2005). FOXP2 and the neuroanatomy of speech and language. *Nature Reviews Neuroscience*, 6: 131–137.

On Von Economo or "spindle" neurons:

Nimchinsky E, Gilissen E, Allman J, Perl D, Erwin J, Hof P. (1999). A neuronal morphologic type unique to humans and great apes, *Proceedings of the National Academy of Sciences*, 96: 5268–5273.

Allman J, Hakeem A, Watson K (2002). Two Phylogenetic Specializations in the human brain. *The Neuroscientist*, 8: 335–346.

On the gap between brain size and human cultural advances:

McBrearty S, Brooks A (2000). The revolution that wasn't: a new interpretation of the origin of modern human behavior. *Journal of Human Evolution*, 39: 453–563.

Diamond J (1992). The Third Chimpanzee: The evolution and future of the human animal. HarperCollins.

Marks J. (2003). What it means to be 98% chimpanzee: Apes, people, and their genes. University of California Press.

On thrombospondins:

Ullian E, Sapperstein S, Christopherson K, Barres B (2001). Control of synapse number by glia. Science, 291: 657–661.

Caceres M, Suwyn C, Maddox M, Thomas J, Preuss T (2007) Increased cortical expression of two synaptogenic thrombospondins in human brain evolution. *Cerebral Cortex* 17:2312–2321.

A discussion of quantity/quality relationships:

Carneiro R. (2000). The transition from quantity to quality: A neglected causal mechanism in accounting for social evolution. *Proceedings of the National Academy of Sciences*, 97: 12926–12931.

It is often argued that environmental variables may have had disproportionate effects on humans. In particular, slightly different environments may have sent one group of humans down a different path, which may have in turn affected the subsequent development of humans across the globe. However much we may like to think that our current human lives were predestined by evolutionary pressure, it's crucial not to overlook external events that may have unexpectedly deflected our evolutionary path. Accidents such as human diseases, or crop diseases, or changing weather patterns, can demonstrably change the course of development of groups of people or animals, and may well have played a role in where we are now.

Gould SJ (1997). Darwinian Fundamentalism. *The New York Review of Books*, 44 (10).

Diamond J (1992). The Third Chimpanzee: The evolution and future of the human animal. HarperCollins.

Weaver T, Roseman C, Stringer C. (2007). Were neandertal and modern human cranial differences produced by natural selection or genetic drift? *Journal of Human Evolution*, 53: 135–145.

Bocquet-Appel J, Demars P, Noiret L, Dobrowsky D. (2005). Estimates of upper palaeolithic metapopulation size in Europe from archaeological data. Journal of Archaeological Science, 32: 1656–1668.

On the nature and evolution of grammars:
Pinker S, Jackendoff R (2004). The faculty of language: What's special about it? *Cognition*, 95: 201–236.

Pinker S. (1999). Words and rules: The ingredients of language. NY: HarperCollins.

Hauser M, Chomsky N, Fitch WT. (2002). The faculty of language: What is it, who has it, and how did it evolve? *Science*, 298: 1569–1579.

Fitch WT, Hauser M (2004) Computational Constraints on Syntactic Processing in a Nonhuman Primate. *Science*, 303: 377–380.

Perruchet P, Rey A (2005) Does the master of center-embedded linguistic structures distinguish humans from nonhuman primates? *Psychonomic Bulletin & Review*, 12: 307–313.

Fisher S, Marcus G (2006) The eloquent ape: genes, brains and the evolution of language. *Nature Reviews Genetics*, 7: 9–20.

Intriguing articles on violence across the primates:
Wrangham R (2004) Killer Species. *Daedalus*, 133: 25–35.

Sapolsky (2006). A natural history of peace. *Foreign Affairs*. 85: 104–120.

CHAPTER 14

Occipital buns are still seen in some subpopulations of humans:
Lieberman DE, Pearson OM, Mowbray KM (2000) Basicranial influence on overall cranial shape. *Journal of Human Evolution*, 38: 291–315.

On the breeding of tame foxes:
Trut L (1999). Early Canid Domestication: The Farm-Fox Experiment. *American Scientist*, 87: 160–169.

A "transcranial magnetic stimulator" (TMS), is an electronic ping-pong paddle that uses a magnetic pulse to briefly and selectively disrupt a particular location in the brain. If you aim it at part of your motor cortex, it can cause you to involuntarily jerk your arm, or your leg, depending on its exact aim. If you aim it at other brain areas, it temporarily disrupts components of thought. It is some-times used in experiments to study the contribution of brain areas to particular tasks; it is even used as an experimental therapeutic device in some cases of schizophrenia and depression, as a far less invasive alternative to some drugs or shocks. Some have reported to unveil abilities ranging from improved drawing, to musical perfect pitch, to enhanced memory—savant abilities like those of Willa, and Les, and Kim from chapter 9. In other words, as we suggested in that chapter, the differences between our abilities and those of savants may be surprisingly slight, so slight that the gap between them may be bridgeable.

Osborne L (2003). Savant for a day. *New York Times Magazine*, June 22.

Scientist Bruce Lahn, among others, has publicly (and very controversially) speculated about the possibility of human speciation:

Regalado A (2006). Scientist's study of brain genes sparks a backlash. *The Wall Street Journal.* Jun 16 2006.

Specific drugs have been shown, in extensive published studies of clinical trials, to selectively enhance the human brain's ability to collect, assemble, and encode information:

Porrino L, Daunais J, Rogers G, Hampson R, Deadwyler S (2005) Facilitation of task performance and removal of the effects of sleep deprivation by an ampakine (CX717) in nonhuman primates. PLoS Biol 3: e299.

Arai A, Kessler M (2007) Pharmacology of ampakine modulators: from AMPA receptors to synapses and behavior. Curr Drug Targets, 8: 583–602.

On the link between Autism and Aspergers, and technical professions such as computer programming and engineering:

Baron-Cohen S, Bolton P, Wheelwright S, Scahill V, Short L, Mead G, Smith A. (1998). Autism occurs more often in families of physicists, engineers, and mathematicians. *Autism*, 2: 296–301.

ACKNOWLEDGMENTS

RHG wishes to thank those who have helped in many ways, including Andrew Felch, Bob Hearn, Yune Lee, Stephanie Gagnon, Peter Tse, Ashok Chandrashekar, Andrew Parker, Travis Green, Cena Miller, Brandyn Webb, James Hughes, Casey Murray, Gabe Weaver, Amy Palmer, Victoria Smith, Jennifer Gaudette, Liya Shuster, Kate Schnippering, Alison Rope, Cynthia Kahlenberg, Meagan Herzog, and Lindsay Zahradka. Thanks to many in the PBS and CS departments. Special thanks to Bill Neukom, Jim Wright, Carol Folt, Janet Terp, Carrie Pelzel, and the Dartmouth administration and development offices. Thanks to Elsa for encouragement and support. Thanks to Rod and Pat for expert advice and counsel. Deep appreciation to Trevor, who suggested many of the directions taken, and to Leann, who was collaborator, craftsman, and consigliere throughout. And inestimable gratitude to Barbara Granger and Dr. Richard Granger. GL thanks Chris Gall: big heart, big brain. Thanks to Luba Ostashevsky at Palgrave Macmillan, an intelligent and patient tutor, and gratitude to Paul Bresnick, who saw the prospect early on, and helped see it all the way through.

BIBLIOGRAPHY

Aboitiz F (1992) The evolutionary origin of the mammalian cerebral cortex. Biological Research 25: 41–49.

Aboitiz F, Morales D, Montiela J. (2003). The evolutionary origin of the mammalian isocortex: Towards an integrated developmental and functional approach. Behavioral and Brain Sciences, 26: 535–586.

Aboitiz F (1993) Further comments on the evolutionary origin of mammalian brain. Med Hypoth 41: 409–418.

Ahmed B, Anderson JC, Martin KAC, Nelson JC (1997) Map of the synapses onto layer 4 basket cells of the primary visual cortex of the cat. J Comp Neurol 380:230–242.

Aleksandrovsky B, Whitson J, Garzotto A, Lynch G, Granger R (1996) An algorithm derived from thalamocortical circuitry stores and retrieves temporal sequences. In: Proceedings of the International Conference on Pattern Recognition 1996, IEEE Computer Society Press, 4: 550–554.

Alexander G, DeLong M. (1985) Microstimulation of the primate neostriatum. I. Physiological properties of striatal microexcitable zones. Journal of Neurophysiology, 53: 1401–1416.

Alexander G, DeLong M. (1985) Microstimulation of the primate neostriatum. II. Somatotopic organization of striatal microexcitable zones and their relation to neuronal response properties. Journal of Neurophysiology, 53: 1417–1430.

Allman J, Hakeem A, Watson K (2002). Two Phylogenetic Specializations in the human brain. *The Neuroscientist*, 8: 335–346.

Ambros-Ingerson J, Granger R, Lynch G (1990) Simulation of paleocortex performs hierarchical clustering. Science 247:1344–1348.

Andre V, Cepeda C, Venegas A, Gomez Y, Levine M. (2006). Altered cortical glutamate receptor function in the R6/2 model of Huntington's Disease. *Journal of Neurophysiology*, 95: 2108–2119.

Anwander A, Tittgemeyer M, von Cramon D, Friederici A, Knosche T (2007). Connectivity-based parcellation of Broca's area. Cerebral Cortex, 17: 816–825.

Bair W, Cavanaugh J, Smith M, Movshon J (2002) The timing of response onset and offset in macaque visual neurons. J Neurosci 22:3189–3205.

Barbas H, Rempel-Clower N (1997) Cortical structure predicts the pattern of corticocortical connections. Cereb Cortex 7:635–646.

Baron-Cohen S, Bolton P, Wheelwright S, Scahill V, Short L, Mead G, Smith A. (1998). Autism occurs more often in families of physicists, engineers, and mathematicians. *Autism*, 2: 296–301.

Barto A. (1995) Adaptive critics and the basal ganglia. In: Models of information processing in the basal ganglia (Ed. Houk J, Davis J, Beiser D) MIT press, pp: 215–232.

Baudry M, Davis J, Thompson R (1999). Advances in synaptic plasticity. MIT Press.

Batardiere A, Barone P, Knoblauch K, Giroud P, Berland M, Dumas A, Kennedy H (2002) Early specification of the hierarchical organization of visual cortical areas in the macaque monkey. Cerebral Cortex 12:453–465.

Bellugi U, Wang P, Jernigan T (1994). Williams syndrome: An unusual neuropsychological profile. In: S.Broman, J.Grafman, eds., Atypical cognitive deficits in developmental disorders. NJ: Erlbaum.

Ben-Shachar M, Dougherty R, Wandell B. (2007). White matter pathways in reading. *Current Opinion in Neurobiology, 17: 1–13.*

Benvenuto J, Jin Y, Casale M, Lynch G, Granger R (2002) Identification of diagnostic evoked response potential segments in Alzheimer's Disease. Experimental Neurology, 176:269–276.

Bhardwaj R, Curtis M, Spalding K, Buchholz B, Fink D, Bjork-Eriksson T, Nordborg C, Gage F, Druid H, Eriksson P, Frise J (2006). Neocortical neurogenesis in humans is restricted to development. *Proceedings of the National Academy of Sciences*, 103: 12564–12568.

Bliss T, Collingridge G (1993). A synaptic model of memory: long-term potentiation in the hippocampus. Nature, 361: 31–39.

Bocquet-Appel J, Demars P, Noiret L, Dobrowsky D. (2005). Estimates of upper palaeolithic metapopulation size in Europe from archaeological data. Journal of Archaeological Science, 32: 1656–1668.

Bourassa J, Deschenes M (1995) Corticothalamic projections from the primary visual cortex in rats: a single fiber study using biocytin as an anterograde tracer. Neuroscience 66:253–263.

Brace, CL (2005). "Race" is a four-letter word: The genesis of the concept. Oxford University Press.

Braitenberg V, Schüz A (1998) Cortex: statistics and geometry of neuronal connectivity, NY: Springer.

Broom R (1918). The Evidence Afforded by the Boskop Skull of a New Species of Primitive Man (*Homo Capensis*). *Anthropological Papers of the American Museum of Natural History*, 23: 65–79.

Brown L, Schneider J, Lidsky T. (1997) Sensory and cognitive functions of the basal ganglia. Current opinion in neurobiology, 7:157–163.

Buonomano D, Merzenich M (1998) Cortical plasticity: from synapses to maps. Annual Review of Neuroscience 21:149–186.

Burkhalter A (1989) Intrinsic connections of rat primary visual cortex: laminar organization of axonal projections. Journal of Comparative Neurology 279: 171–186.

Bush P, Sejnowski T (1996) Inhibition synchronizes sparsely connected cortical neurons within and between columns in realistic network models. Journal of Computational Neuroscience 3:91–110.

Buxhoeveden D, Switala AE, Litaker M, Roy E, Casanova M. (2001) Lateralization in human planum temporale is absent in nonhuman primates. *Brain Behavior and Evolution*, 57: 349–358.

Buxhoeveden D, Switala A, Roy E, Litaker M, Casanova M. (2001) Morphological differences between minicolumns in human and non human primate cortex. *American Journal of Physical Anthropology*, 115: 361–371.

Caceres M, Suwyn C, Maddox M, Thomas J, Preuss T (2007) Increased cortical expression of two synaptogenic thrombospondins in human brain evolution. *Cerebral Cortex* 17:2312–2321.

Canales J, Capper-Loup C, Hu D, Choe E, Upadhyay U, Graybiel A (2002) Shifts in striatal responsivity evoked by chronic stimulation of dopamine and glutamate systems. Brain 125:2353–2363.

Cantalupo C, Hopkins W (2001). Asymmetric Broca's area in great apes. Nature, 414: 505.

Capek K (1920) RUR: Rossum's Universal Robots (Rossumovi Univerzalni Roboti).

Caplan J, Madsen, J., Raghawachari, S., Kahana, M. (2001) Distinct patterns of brain oscillations underlie two basic parameters of human maze learning. Journal of Neurophysiology 86:368–380.

Carneiro R. (2000). The transition from quantity to quality: A neglected causal mechanism in accounting for social evolution. *Proceedings of the National Academy of Sciences*, 97: 12926–12931.

Carroll S. (2005). Endless forms most beautiful. The new science of evo devo and the making of the animal kingdom. NY: WW Norton.

Castro-Alamancos M, Connors B (1997) Thalamocortical synapses. Prog Neurobiol 51:581–606.

Castro-Alamancos M, Donoghue J, Connors B (1995) Different forms of synaptic plasticity in somatosensory and motor areas of the neocortex. J Neurosci 15:5324–5333.

Changizi M (2003) The brain from 25,000 feet. Springer.

Charpier S, Deniau J. (1997) In vivo activity-dependent plasticity at the cortico-striatal connections: Evidence for physiological long-term potentiation. Proceedings of the National Academy of Sciences, 94: 7036–70340.

Chrobak J, Buzsaki G (1998) Gamma oscillations in entorhinal cortex of the freely behaving rat. Journal of Neuroscience 18: 388–398.

Chesselet M, Delfs J. (1996) Basal ganglia and movement disorders: update. Trends in Neuroscience, 19: 417–422.

Churchland P, Sejnowski T (1992) The computational brain. MIT Press.

Cohen A, Rossignol S, Grillner S (1988). Neural control of rhythmic movements in vertebrates. Wiley & Sons.

Conley M, Diamond IT (1990) Organization of the Visual Sector of the Thalamic Reticular Nucleus in Galago. European Journal of Neuroscience 2:211–226.

Coultrip R, Granger R (1994) LTP learning rules in sparse networks approximate Bayes classifiers via Parzen's method. Neural Networks 7: 463–476.

Coultrip R, Granger R, Lynch G (1992) A cortical model of winner-take-all competition via lateral inhibition. Neural Networks 5: 47–54.

Cox C, Huguenard J, Prince D (1997) Nucleus reticularis neurons mediate diverse inhibitory effects in thalamus. Proceedings of the National Academy of Sciences, 94: 8854–8859.

Creutzfeldt O, Nothdurft H (1978) Representation of complex visual stimuli in the brain. Naturwissenschaften 65: 307–318.

Crichton M. (1972). The Terminal Man. A.Knopf.

Critchley M, Critchley E (1999). John Hughlings Jackson. Oxford University Press.

Dart R (1923). Boskop remains from the south-east African coast. *Nature*, 112: 623–625.

Dart R (1940). Recent discoveries bearing on human history in southern Africa. *Journal of the Royal Anthropological Institute of Great Britain and Ireland*, 70: 13–27.

Day M, Langston R, Morris R (2003). Glutamate receptor mediated encoding and retrieval of paired-associate learning. Nature, 424: 205–209.

DeFelipe J, Jones E (1991) Parvalbumin immunoreactivity reveals layer IV of monkey cerebral cortex as a mosaic of microzones of thalamic afferent terminations. Brain Research 562: 39–47.

Delgado J (1969) Physical control of the mind: Toward a pscyhocivilized society. Harper & Row.

Dennett D (1996) Darwin's dangerous idea. Simon & Schuster.

Deschenes M, Veinante P, Zhang Z (1998) The organization of corticothalamic projections: reciprocity versus parity. Brain Research Reviews 28: 286–308.

Destexhe A, Contreras D, Steriade M (1999) Cortically-induced coherence of a thalamic-generated oscillation. Neuroscience 92: 427–443.

Deutsch GK, Dougherty RF, Bammer R, Siok WT, Gabrieli JD, Wandell B (2005). Children's reading performance is correlated with white matter structure measured by diffusion tensor imaging. *Cortex*, 41: 354–363.

Diamond J (1992). The Third Chimpanzee: The evolution and future of the human animal. HarperCollins.

Diamond M, Armstrong-James M, Ebner F (1992) Somatic sensory responses in the rostral sector of the posterior group (POm) and in the ventral posterior medial nucleus (VPM) of the rat thalamus. Journal of Comparative Neurology 318: 462–476.

Diamond M, Armstrong-James M, Budway M, Ebner F (1992) Somatic sensory responses in the rostral sector of the posterior group (POm) and the ventral posterior medial nucleus (VPM) of the rat thalamus: dependence on the barrel field cortex. Journal of Comparative Neurology 319: 66–84.

Dorus S, Vallender EJ, Evans PD, Anderson JR, Gilbert SL, Mahowald M, Wyckoff GJ, Malcom CM, Lahn BT. (2004). Accelerated evolution of nervous system genes in the origin of *Homo sapiens. Cell*, 119:1027.

Douglas R, Mahowald M, Martin K, Stratford K (1996). The role of synapses in cortical computation. Journal of Neurocytology, 25: 893–911.

Drennan M (1931). Pedomorphism in the pre-bushman skull. American Journal of Physical Anthropology, 16: 203–210.

Eiseley L. (1958) The Immense Journey. London: V.Gollancz.

Eldredge N, Gould S. (1972) Punctuated equilibria: an alternative to phyletic gradualism. In: *Models In Paleobiology* (Ed. by T. J. M. Schopf). Freeman Cooper.

Enard W, Paabo S. (2004). Comparative primate genomics. *Annual Reviews: Genomics and Human Genetics*, 5: 351–378.

Evans P, Gilbert S, Mekel-Bobrov N, Vallender E, Anderson J, Vaez-Azizi L, Tishkoff S, Hudson R, Lahn B. (2005). Microcephalin, a gene regulating brain size, continues to evolve adaptively in humans. Science, 309: 1717–1720.

Felch A, Granger R (2007). The hypergeometric connectivity hypothesis: Divergent performance of brain circuits with different synaptic connectivity distributions. Brain Research, doi: 10.1016/j.brainres.2007.06.04

Ferrier D (1876) Functions of the brain. London: Smith, Elder.

Finlay B, Darlington R (1995). Linked regularities in the development and evolution of mammalian brains. *Science*, 268: 1578–1584.

Finnerty G, Roberts L, Connors B. (1999). Sensory experience modifies the short-term dynamics of neocortical synapses. Nature, 400: 367–371.

Fisher S, Marcus G (2006) The eloquent ape: genes, brains and the evolution of language. *Nature Reviews Genetics*, 7: 9–20.

Fitch T, Hauser M (2004) Computational constraints on syntactic processing in a nonhuman primate. Science 303: 377–380.

Fitzpatrick D, Lund JS, Schmechel DE, Towles AC (1987) Distribution of Gabaergic Neurons and Axon Terminals in the Macaque Striate Cortex. Journal of Comparative Neurology, 264: 73–91.

FitzSimons FW (1915). Palaeolithic man in South Africa. *Nature*, 95: 615–616.

Freedman D, Riesenhuber M, Poggio T, Miller E (2001) Categorical representation of visual stimuli in the primate prefrontal cortex. Science, 291: 312–316.

Freedman D, Riesenhuber M, Poggio T, Miller E (2002). Visual categorization and the primate prefrontal cortex: Neurophysiology and behavior. Journal of Neurophysiology, 88: 929–941.

Freund T, Martin K, Soltesz I, Somogyi P, Whitteridge D (1989) Arborisation pattern and postsynaptic targets of physiologically identified thalamocortical afferents in striate cortex of the macaque monkey. Journal of Comparative Neurology, 289: 315–336.

Furlong J, Felch A, Nageswaran J, Dutt N, Nicolau A, Veidenbaum A, Chandreshekar A, Granger R. (2007). Novel brain-derived algorithms scale linearly with number of processing elements. Proceedings of the International Conference on Parallel Computing, (parco.org) 2007.

Gage F (2002) Neurogenesis in the adult brain. *Journal of Neuroscience* 22: 612–613.

Gardiner G, Currant A (1996) The Piltdown Hoax: Who done it? Linnean Society of London.

Gardner H. (1993). Frames of Mind: The theory of multiple intelligences. Basic Books.

Galloway A (1937). The Characteristics of the Skull of the Boskop Physical Type. *American Journal of Physical Anthropology*, 32: 31–47.

Galuske RA, Schlote W, Bratzke H, Singer W (2000) Interhemispheric asymmetries of the modular structure in human temporal cortex. Science 289: 1946–1949.

Gazzaniga M (2000) Regional differences in cortical organization. Science 289: 1887–1888.

Gazzaniga M (2004) The cognitive neurosciences. MIT Press.

Gerfen, C. (1992) The neostriatal mosaic: multiple levels of compartmental organization in the basal ganglia. Annual Review of Neuroscience, 15: 285–320.

Gluck M, Granger R (1993) Computational models of neural bases of learning & memory. Annual Review of Neuroscience 16: 667–706.

Golarai1 G, Ghahremani1 D, Whitfield-Gabrieli S, Reiss A, Eberhardt J, Gabrieli J, Grill-Spector K (2007). Differential development of high-level visual cortex correlates with category-specific recognition memory. *Nature Neuroscience* 10: 512–522.

Gould E, Reeves A, Graziano S, Gross C. (1999). Neurogenesis in the Neocortex of Adult Primates. *Science*, 286: 548–552.

Gould SJ (1990). Wonderful life: The Burgess Shale and the nature of history. NY: WW Norton.

Gould SJ (1997). Darwinian Fundamentalism. *The New York Review of Books*, 44 (10).

Granger R, Lynch G (1991) Higher olfactory processes: perceptual learning and memory. Current Opinion in Neurobiology 1: 209–214.

Granger R, Staubli U, Powers H, Otto T, Ambros-Ingerson J, Lynch G. (1991). Behavioral tests of a prediction from a cortical network simulation. *Psychological Science*, 2: 116–118.

Granger R, Whitson J, Larson J, Lynch G (1994) Non-Hebbian properties of long-term potentiation enable high-capacity encoding of temporal sequences. Proceedings of the National Academy of Sciences, 91: 10104–10108.

Granger R, Wiebe S, Taketani M, Ambros-Ingerson J, Lynch G (1997) Distinct memory circuits comprising the hippocampal region. Hippocampus 6: 567–578.

Granger R. (2005) Brain circuit implementation: High-precision computation from low-precision components. In: Replacement Parts for the Brain (T.Berger, D.Glanzman, Eds) MIT Press., 277–294.

Granger R. (2006) Engines of the brain: The computational instruction set of human cognition. AI Magazine, 27: 15–32.

Granger R. (2006) The evolution of computation in brain circuitry. Behavioral and Brain Sciences, 29: 17–18.

Graybiel A. (1995) Building action repertoires: memory and learning functions of the basal ganglia. Current Opinion in Neurobiology, 5:733–741.

Graybiel A. (1997) The basal ganglia and cognitive pattern generators. Schizophrenia bulletin, 23: 459–69.

Grill-Spector K, Kushnir T, Hendler T, Malach R. (2000). The dynamics of object-selective activation correlate with recognition performance in humans. Nature neuroscience, 3: 837–843.

Grill-Spector K (2003). The neural basis of object perception. Current Opinion in Neurobiology, 13: 1–8.

Grill-Spector K, Kanwisher N (2005). Visual recognition: As soon as you know it is there, you know what it is. Psychological Science, 16: 152–160.

Grill-Spector K, Sayres R, Ress D (2006). High-resolution imaging reveals highly selective nonface clusters in the fusiform face area. Nature neuroscience, 9: 1177–1185.

Grossberg S (1976) Adaptive pattern classification and universal recoding. Biological Cybernetics 23:121–134.

Guido W, Lu S, Sherman S (1992) Relative contributions of burst and tonic responses to receptive field properties of lateral geniculate neurons in the cat. Journal of Neurophysiology 68:2199–2211.

Guido W, Lu S, Vaughan J, Godwin D, Sherman S (1995) Receiver operating characteristic (ROC) analysis of neurons in the cat's lateral geniculate nucleus during tonic and burst response mode. Visual Neuroscience 12: 723–741.

Gullapalli V, Franklin J, Benbrahim H. (1994) Acquiring robot skills via reinforcement learning. IEEE Control Systems Magazine, 14(1):13–24.

Haughton S (1917). Preliminary note on the ancient human skull remains from the Transvaal. *Transactions of the Royal Society of South Africa*, 6: 1–14.

Hauser M, Chomsky N, Fitch WT. (2002). The faculty of language: What is it, who has it, and how did it evolve? *Science*, 298: 1569–1579.

Hawking S (2001). The Universe in a Nutshell. Bantam.

Hawkins J, Blakeslee S (2004) On intelligence. Times Books.

Haxby J (2006). Fine structure in representations of faces and objects. *Nature Neuroscience* 9: 1084–1086.

Henry G (1991) Afferent inputs, receptive field properties and morphological cell types in different layers. In: Vision and visual dysfunction. (Leventhal A, ed), pp. 223–240. London: Macmillan Press.

Hauser M, Weiss D, Marcus G (2002). Rule learning by cotton-top tamarins. Cognition, 86: 815=822.

Hauser M, Chomsky N, Fitch T (2002). The language faculty: What is it, who has it, and how did it evolve? Science 298: 1569–1579.

Herkenham M (1980) Laminar organization of thalamic projections to the rat neocortex. Science 207: 532–535.

Herkenham M (1986) New perspesties on the organization and evolution of nonspecific thalamocortical projections. In: Cerebral Cortex (Jones EG, Peters, A., ed). New York: Plenum Press.

Hess G, Aizenman CD, Donoghue JP (1996) Conditions for the induction of long-term potentiation in layer II/III horizontal connections of the rat motor cortex. Journal of Neurophysiology 75: 1765–1778.

Heynen AJ, Bear MF (2001) Long-term potentiation of thalamocortical transmission in the adult visual cortex in vivo. Journal of Neuroscience 21:9801–9813.

Hirsch J, Crepel F (1990) Use-dependent changes in synaptic efficacy in rat prefrontal neurons in vitro. Journal of Physiology 427: 31–49.

Hobolth A, Christensen OF, Mailund T, Schierup MH (2007) Genomic Relationships and Speciation Times of Human, Chimpanzee, and Gorilla Inferred from a Coalescent Hidden Markov Model. *PLoS Genetics* 3(2): e7 doi:10.1371/journal.pgen.0030007

Holloway R (2002). Brief communication: How much larger is the relative volume of area 10 of the prefrontal cortex in humans? *American Journal of Physical Anthropology*, 118: 399–401.

Hopkins W, Cantalupo C, Taglialatela J (2006). Handedness is associated with asymmetries in gyrification of the cerebral cortex of chimpanzees. Cerebral Cortex, 17: 1750–1756.

Houk J, Wise S (1995). Distributed modular architectures linking basal ganglia, cerebellum, and cerebral cortex: their role in planning and controlling action. Cerebral Cortex, 5: 95–110.

Huang CL, Winer JA (2000) Auditory thalamocortical projections in the cat: laminar and areal patterns of input. Journal of Comparative Neurology, 427: 302–331.

Hubel D, Wiesel T (1977) Functional architecture of macaque monkey visual cortex. Proceedings of the Royal Society of London Biological Sciences 198: 1–59.

Huguenard JR, Prince DA (1994) Clonazepam suppresses GABAB-mediated inhibition in thalamic relay neurons through effects in nucleus reticularis. Journal of Neurophysiology 71:2576–2581.

Hung C, Kreiman G, Poggio T, DiCarlo J (2005). Fast readout of object identity from macaque inferior temporal cortex. Science, 310: 863–866.

Ichinohe N, Rockland KS (2002) Parvalbumin positive dendrites co-localize with apical dendritic bundles in rat retrosplenial cortex. Neuroreport 13:757–761.

Iriki A, Pavlides C, Keller A, Asanuma H (1991) Long-term potentiation of thalamic input to the motor cortex induced by coactivation of thalamocortical and corticocortical afferents. Journal of Neurophysiology 65: 1435–1441.

Jackson JH (1925) Neurological fragments. London: Oxford University Press

James W (1890). The Principles of Psychology. Boston: Henry Holt. (p.462)

Jensen KF, Killackey HP (1987) Terminal arbors of axons projecting to the somatosensory cortex of the adult rat. I. The normal morphology of specific thalamocortical afferents. Journal of Neuroscience 7:3529–3543.

Jerison H. (1974) Evolution of the Brain and Intelligence. Academic Press.

Johnson J, Olshausen B (2003) Timecourse of neural signatures of object recognition. Journal of Vision, 3: 499–512.

Johnson S (2004) Mind wide open. Scribner.

Jolicoeur P, Gluck M, Kosslyn SM (1984) Pictures and names: making the connection. Cognitive Psychology 16:243–275.

Jones E (1981) Functional subdivision and synaptic organization of the mammalian thalamus. International Review of Physiology 25: 173–245.

Jones E (1998) A new view of specific and nonspecific thalamocortical connections. Advances in Neurology 77:49–71.

Jones E (2001) The thalamic matrix and thalamocortical synchrony. Trends in Neuroscience 24:595–601.

Kandel E, Schwartz J, Jessell T (2000). Principles of neural science, 4th ed., NY: McGraw-Hill.

Karmiloff-Smith A (1998). Development itself is the key to understanding developmental disorders. *Trends in Cognitive Sciences*, 2: 389–398.

Keith A (1914) The antiquity of man. London: Williams and Norgate.

Keller A, White EL (1989) Triads: a synaptic network component in the cerebral cortex. Brain Research 496: 105–112.

Kelly JP, Wong D (1981) Laminar connections of the cat's auditory cortex. Brain Research 212: 1–15.

Kenan-Vaknin G, Teyler TJ (1994) Laminar pattern of synaptic activity in rat primary visual cortex: comparison of in vivo and in vitro studies employing the current source density analysis. Brain Research 635: 37–48.

Kilborn K, Granger, R, Lynch, G (1996) Effects of LTP on response selectivity of simulated cortical neurons. Journal of Cognitive Neuroscience 8: 338–353.

Killackey H, Ebner F (1973) Convergent projection of three separate thalamic nuclei on to a single cortical area. Science 179: 283–285.

Killackey HP, Ebner FF (1972) Two different types of thalamocortical projections to a single cortical area in mammals. Brain Behavior and Evolution 6:141–169.

Kim HG, Fox K, Connors BW (1995) Properties of excitatory synaptic events in neurons of primary somatosensory cortex of neonatal rats. Cerebral Cortex 5:148–157.

Kimura A, Caria MA, Melis F, Asanuma H (1994) Long-term potentiation within the cat motor cortex. Neuroreport 5:2372–2376.

Kirkwood A, Dudek SM, Gold JT, Aizenman CD, Bear MF (1993) Common forms of synaptic plasticity in the hippocampus and neocortex in vitro. Science 260:1518–1521.

Klingberg T, Hedehus M, Temple E, Salz T, Gabrieli J, Moseley M, Poldrack R (2000). Microstructure of temporo-parietal white matter as a basis for reading ability: Evidence from diffusion tensor magnetic resonance imaging. *Neuron*, 25: 493–500.

Komatsu Y, Fujii K, Maeda J, Sakaguchi H, Toyama K (1988) Long-term potentiation of synaptic transmission in kitten visual cortex. Journal of Neurophysiology 59:124–141.

Kornack D, Rakic P (2000). Cell proliferation without neurogenesis in adult primate neocortex. *Science*, 294: 2127–2130.

Kosslyn S, Thompson W, Ganis G (2006). The case for mental imagery. Oxford University Press.

Kudoh M, Shibuki K (1996) Long-term potentiation of supragranular pyramidal outputs in the rat auditory cortex. Experimental Brain Research 110:21–27.

Kuhl P, Tsao, F., Zhang, Y., DeBoer, B. (2001) Language, culture, mind, brain: Progress at the margins between disciplines. Annals of the New York Academy of Science, 935:136–174.

Kuroda M, Yokofujita J, Murakami K (1998) An ultrastructural study of the neural circuit between prefrontal cortex and the mediodorsal nucleus of the thalamus. Progress in Neurobiology 54:417–458.

Lai C, Fisher S, Hurst J, Levy E, Hodgson S, Fox M, Jeremiah S, Povey S, Jamison D, Green E, Vargha-Khadem F, Monaco A (2000). The SPCH1 region on human 7q31: Genomic characterization of the critical interval and localization of translocations associated with speech and language disorder. *American Journal of Human Genetics*, 67: 357–368.

LeDoux J (2002) Synaptic self: How our brains become who we are. Viking.

Levin D, Simons D (1997) Failure to detect changes to attended objects in motion pictures, *Psychonomic Bulletin and Review* 4: 501–506.

Lieberman DE, Pearson OM, Mowbray KM (2000) Basicranial influence on overall cranial shape. *Journal of Human Evolution*, 38: 291–315.

Linke R, Schwegler H (2000) Convergent and complementary projections of the caudal paralaminar thalamic nuclei to rat temporal and insular cortex. Cerebral Cortex 10:753–771.

Liu J, Harris A, Kanwisher N (2002) Stages of processing in face perception: an MEG study. Nature neuroscience, 5: 910–916.

Liu XB, Jones EG (1999) Predominance of corticothalamic synaptic inputs to thalamic reticular nucleus neurons in the rat. Journal of Comparative Neurology 414:67–79.

Lorente de No R (1938) Cerebral cortex: Architecture, intracortical connections, motor projections. In: Physiology of the nervous system (Fulton J, ed), pp. 291–340. London: Oxford University Press.

Lynch G (1986). Synapses, circuits and the beginnings of memory. MIT Press.

Lynch G, Hechtel S, Jacobs D (1983). Neonate size and evolution of brain size in the anthropoid primates. J. Human Evolution, 12: 519–522.

Macrides F (1975) Temporal relationships between hippocampal slow waves and exploratory sniffing in hamsters. Behavior Biology 14:295–308.

Macrides F, Eichenbaum HB, Forbes WB (1982) Temporal relationship between sniffing and the limbic theta rhythm during odor discrimination reversal learning. Journal of Neuroscience 2:1705–1717.

Magee JC (2000) Dendritic integration of excitatory synaptic input. Nature Reviews Neuroscience 1:181–190.

Magee JC, Cook EP (2000) Somatic EPSP amplitude is independent of synapse location in hippocampal pyramidal neurons. Nature Neuroscience 3:895–903.

Marcus G. (2004). The birth of the mind: How a tiny number of genes creates the complexities of human thought. NY: Basic Books.

Marino L, Uhen M, Pyenson N, Frohlich B (2003). Reconstructing cetacean brain evolution using computed tomography. *The Anatomical Record.* 272B: 107–117.

Marks J. (2003). What it means to be 98% chimpanzee: Apes, people, and their genes. University of California Press.

McBrearty S, Brooks A (2000). The revolution that wasn't: a new interpretation of the origin of modern human behavior. *Journal of Human Evolution*, 39: 453–563.

McCollum J, Larson J, Otto T, Schottler F, Granger R, Lynch G (1991) Short-latency single-unit processing in olfactory cortex. Journal of Cognitive Neuroscience 3:293–299.

McCormick DA, Feeser HR (1990) Functional implications of burst firing and single spike activity in lateral geniculate relay neurons. Neuroscience 39:103–113.

McCormick D, Bal T (1994) Sensory gating mechanisms of the thalamus. Current Opinion in Neurobiology 4: 550–556.

McHenry H (1992). Body size and proportions in early hominids. American Journal of Physical Anthropology, 87: 407–431.

McHenry H (1994). Tempo and mode in human evolution. *Proceedings of the National Academy of Sciences*, 91: 6780–6786.

Mekel-Bobrov N, Gilbert S, Evans PD, Vallender E, Anderson J, Hudson R, Tishkoff S, Lahn B. (2005) Ongoing adaptive evolution of *ASPM*, a brain size determinant in *Homo sapiens. Science*, 309: 1720.

Mervis C, Rosch, E. (1981) Categorization of natural objects. Annual Review of Psychology 32:293–299.

Mitani A, Shimokouchi M, Itoh K, Nomura S, Kudo M, Mizuno N (1985) Morphology and laminar organization of electrophysiologically identified neurons in primary auditory cortex in the cat. Journal of Comparative Neurology 235: 430–447.

Mitchell BD, Cauller LJ (2001) Corticocortical and thalamocortical projections to layer I of the frontal neocortex in rats. Brain Research 921:68–77.

Molinari M, Dell'Anna ME, Rausell E, Leggio MG, Hashikawa T, Jones EG (1995) Auditory thalamocortical pathways defined in monkeys by calcium-binding protein immunoreactivity. Journal of Comparative Neurology, 362:171–194.

Moorkanikara J, Chandrashekar A, Felch A, Furlong J, Dutt N, Nicolau A, Veidenbaum A, Granger R (2007). Accelerating brain circuit simulations of object recognition with a Sony PlayStation 3. International Workshop on Innovative Architectures (IWIA).

Mountcastle VB (1957) Modality and topographic properties of single neurons of cat's somatic sensory cortex. Journal of Neurophysiology 20:408–434.

Mountcastle VB (1978) Brain mechanisms for directed attention. Journal of the Royal Society for Medicine 71:14–28.

Mukherjee P, Kaplan E (1995) Dynamics of neurons in the cat lateral geniculate nucleus: in vivo electrophysiology and computational modeling. Journal of Neurophysiology 74:1222–1243.

Mumford D (1992) On the computational architecture of the neocortex. II. The role of cortico-cortical loops. Biological Cybernetics 66:241–251.

Murray E, Bussey T (1999) Perceptual-mnemonic functions of the perirhinal cortex. Trends in cognitive sciences, 3: 142–151.

Nimchinsky E, Gilissen E, Allman J, Perl D, Erwin J, Hof P. (1999). A neuronal morphologic type unique to humans and great apes, *Proceedings of the National Academy of Sciences*, 96: 5268–5273.

Nowlan S & Sejnowski T (1995) A selection model for motion processing in area MT of primates. Journal of Neuroscience, 15: 1195–1214.

O'Donnell T, Hauser M, Fitch T (2004). Using mathematical models of language experimentally. Trends in Cognitive Science, 9: 284–289.

O'Kusky J, Colonnier M (1982) A laminar analysis of the number of neurons, glia, and synapses in the adult cortex (area 17) of adult macaque monkeys. Journal of Comparative Neurology, 210:278–290.

Olshausen B, Field D (1996). Emergence of simple-cell receptive field properties by learning a sparse code for natural images. Nature, 381: 607–609.

Olson CR, Musil SY (1992) Topographic organization of cortical and subcortical projections to posterior cingulate cortex in the cat: evidence for somatic, ocular, and complex subregions. Journal of Comparative Neurology, 324:237–260.

Osborne L (2003). Savant for a day. *New York Times Magazine,* June 22.

Patterson N, Richter DJ, Gnerre S, Lander ES, Reich D (2006) Genetic evidence for complex speciation of humans and chimpanzees. *Nature* 441: 1103–1108.

Perruchet P, Rey A (2005) Does the master of center-embedded linguistic structures distinguish humans from nonhuman primates? *Psychonomic Bulletin & Review,* 12: 307–313.

Pesaran B, Pezaris J, Sahani M, Mitra P, Andersen R (2002) Temporal structure in neuronal activity during working memory in macaque parietal cortex. Nature Neuroscience 5:805–811.

Peters A, Payne B (1993) Numerical Relationships between Geniculocortical Afferents and Pyramidal Cell Modules in Cat Primary Visual-Cortex. Cerebral Cortex 3:69–78.

Peters A, Payne B, Budd J (1994) A Numerical-Analysis of the Geniculocortical Input to Striate Cortex in the Monkey. Cerebral Cortex 4:215–229.

Peterson BE, Goldreich D, Merzenich MM (1998) Optical imaging and electrophysiology of rat barrel cortex. I. Responses to small single-vibrissa deflections. Cerebral Cortex 8:173–183.

Pinker S (1999). Words and rules: the ingredients of language. New York: HarperCollins.

Pinker S, Jackendoff R (2004). The faculty of language: What's special about it? *Cognition*, 95: 201–236.

Porrino L, Daunais J, Rogers G, Hampson R, Deadwyler S (2005) Facilitation of task performance and removal of the effects of sleep deprivation by an ampakine (CX717) in nonhuman primates. PLoS Biol 3: e299.

Preuss T (1995). Do rats have prefrontal cortex? The Rose-Woolsey-Akert program reconsidered. Journal of Cognitive Neuroscience, 7: 1–24.

Preuss T, Qi H, Kaas J. (1999) Distinctive compartmental organization of human primary visual cortex. *Proceedings of the National Academy of Sciences*, 96: 11601–11606.

Preuss T (2000). What's human about the human brain? In: The New Cognitive Neurosciences. M.Gazzaniga (Ed.), Cambridge, MA: MIT Press, pp.1219–1234.

Pycraft W (1925). One the Calvaria Found at Boskop, Transvaal, in 1913, and Its Relationship to Cromagnard and Negroid Skulls. *Journal of the Royal Anthropological Institute of Great Britain and Ireland*, 55: 179–198.

Ramus F, Hauser M, Miller C, Morris D, Mehler J (2000). Language discrimination by human newborns and by cotton-top tamarin monkeys. Science, 288: 349–351.

Rauschecker JP, Tian B, Pons T, Mishkin M (1997) Serial and parallel processing in rhesus monkey auditory cortex. Journal of Comparative Neurology, 382:89–103.

Read HL, Winer JA, Schreiner CE (2002) Functional architecture of auditory cortex. Current Opinion in Neurobiology, 12:433–440.

Reber P, Squire L (1998) Encapsulation of implicit and explicit memory in sequence learning. Journal of Cognitive Neuroscience 10:248–263.

Reber PJ, Stark CE, Squire LR (1998a) Cortical areas supporting category learning identified using functional MRI. Proceedings of the National Academy of Sciences, 95:747–750.

Reber PJ, Stark CE, Squire LR (1998b) Contrasting cortical activity associated with category memory and recognition memory. Learning and Memory, 5:420–428.

Recanzone G, Guard D, Phan M (2000) Frequency and intensity response properties of single neurons in auditory cortex of the behaving macaque monkey. Journal of Neurophysiology, 83:2315–2331.

Reddy L, Kanwisher N (2006). Coding of visual objects in the ventral stream. *Current Opinion in Neurobiology* 16: 408–414.

Redon R, Ishikawa S, Fitch K, Feuk L, and 39 additional authors (2006). Global variation in copy number in the human genome. *Nature*, 444: 444–454.

Reep RL, Corwin JV (1999) Topographic organization of the striatal and thalamic connections of rat medical agranular cortex. Brain Research, 841:43–52.

Regalado A (2006). Scientist's study of brain genes sparks a backlash. *The Wall Street Journal.* Jun 16, 2006.

Reinagel P, Godwin D, Sherman SM, Koch C (1999) Encoding of visual information by LGN bursts. Journal of Neurophysiology, 81:2558–2569.

Reiner A, Perkel D, Bruce L, Butler A, Csillag A, Kuenzel W, Medina L, Paxinos G, Shimizu T, Striedter G, Wild M, Ball G, Powers A, White S, Hough G, Kubikova L, Smulders T, Wada K, Dugas J, Husband S, Yamamoto K, Yu J, Siang C, Jarvis E. (2004). Revised nomenclature for avian telencephalon and some related brainstem nuclei. Journal of Comparative Neurology, 473: 377–414.

Ribak CE, Peters A (1975) An autoradiographic study of the projections from the lateral geniculate body of the rat. Brain Research 92:341–368.

Rieck RW, Carey RG (1985) Organization of the rostral thalamus in the rat: evidence for connections to layer I of visual cortex. Journal of Comparative Neurology, 234:137–154.

Rioult-Pedotti MS, Friedman D, Donoghue JP (2000) Learning-induced LTP in neocortex. Science 290:533–536.

Rockel AJ, Hiorns RW, Powell TP (1980) The basic uniformity in structure of the neocortex. Brain 103:221–244.

Rockland KS (2002) Visual cortical organization at the single axon level: a beginning. Neuroscience Research 42:155–166.

Rodriguez A, Whitson J, Granger R (2004) Derivation & analysis of basic computational operations of thalamocortical circuits. Journal of Cognitive Neuroscience, 16: 856–877.

Roe A, Pallas S, Kwon Y, Sur M. (1992) Visual projections routed to the auditory pathway in ferrets: receptive fields of visual neurons in primary auditory cortex. *Journal of Neuroscience* 12: 3651–3664.

Romo R, Hernandez A, Zainos A (2004). Neuronal correlates of a perceptual decision in ventral premotor cortex. Neuron, 41: 165–173.

Rouiller EM, Welker E (1991) Morphology of corticothalamic terminals arising from the auditory cortex of the rat: a Phaseolus vulgaris-leucoagglutinin (PHA-L) tracing study. Hearing Research 56:179–190.

Rouiller EM, Liang FY, Moret V, Wiesendanger M (1991) Patterns of Corticothalamic Terminations Following Injection of Phaseolus-Vulgaris Leukoagglutinin (Pha-L) in the Sensorimotor Cortex of the Rat. Neuroscience Letters 125:93–97.

Rumelhart D, Zipser, D (1985) Feature discovery by competitive learning. Cognitive Science 9:75–112.

Ryugo DK, Killackey HP (1974) Differential telencephalic projections of the medial and ventral divisions of the medial geniculate body of the rat. Brain Research 82:173–177.

Saleem K, Suzuki W, Tanaka K, Hashikawa T (2000) Connections between anterior inferotemporal cortex and superior temporal sulcus regions in macaque monkey. Journal of Neuroscience 20:5083–5101.

Sapolsky (2006). A natural history of peace. *Foreign Affairs.* 85: 104–120.

Schack B, Klimesch W (2002) Frequency characteristics of evoked and oscillatory electroencephalic activity in a human memory scanning task. Neuroscience Letters 331:107.

Scheel M (1988) Topographic organization of the auditory thalamocortical system in the albino rat. Anatomy and Embryology (Berl) 179:181–190.

Schlaghecken F (1998) On processing "beasts" and "birds": An event-related potential study on the representation of taxonomic structure. Brain and Language 64.

Schwartz, J., Tattersall, I. (2003). The Human Fossil Record, Vols 1–4. Wiley.

Schwartz J (2006) Race and the odd history of human paleontology. The Anatomical Record, 289B: 225–240.

Scoville W, Milner B (1957). Loss of recent memory after bilateral hippocampal lesions. Journal of Neurology, Neurosurgery, and Psychiatry, 20: 11–21.

Semendeferi K, Damasio H, Frank R, Van Hoesen G (1997). The evolution of the frontal lobes: a volumetric analysis based on three-dimensional reconstruction of magnetic resonance scans of human and ape brains. Journal of Human Evolution, 32: 375–388.

Semendeferi K, Armstrong E, Schleicher A, Zilles K, Van Hoesen G (2001). Prefrontal cortex in humans and apes: A comparative study of area 10. American Journal of Physical Anthropology, 114: 224–241.

Semendeferi K, Lu A, Schenker N, Damasio H (2002). Humans and great apes share a large frontal cortex. *Nature Neuroscience*, 5: 272–276.

Shimono K, Brucher F, Granger R, Lynch G, Taketani M (2000) Origins and distribution of cholinergically induced beta rhythms in hippocampal slices. Journal of Neuroscience 20:8462–8473.

Shipman P (2002) The man who found the missing link: Eugene Dubois and his lifelong quest to prove Darwin right. Harvard University Press.

Shultz W. (1997) Dopamine neurons and their role in reward mechanisms. Current opinion in neurobiology, 7:191–197.

Shultz W, Dayan P, Montague P. (1997) A neural substrate of prediction and reward. Science, 275:1593–1599.

Schwark HD, Jones EG (1989) The distribution of intrinsic cortical axons in area 3b of cat primary somatosensory cortex. Experimental Brain Research, 78:501–513.

Seki K, Kudoh M, Shibuki K (2001) Sequence dependence of post-tetanic potentiation after sequential heterosynaptic stimulation in the rat auditory cortex. Journal of Physiology, 533:503–518.

Sherman SM (2001) Tonic and burst firing: dual modes of thalamocortical relay. Trends in Neuroscience 24:122–126.

Sherrington C (1906). The integrative action of the nervous system. NY: Scribner.

Sherwood C, Broadfield D, Holloway R, Gannon P, Hof P. (2003). Variability of Broca's area homologue in African great apes: Implications for language evolution. *The Anatomical Record*. 271A: 276–285.

Silberberg G, Gupta A, Markram H (2002) Stereotypy in neocortical microcircuits. Trends in Neuroscience 25:227–230.

Sobotka S, Ringo, J. (1997) Saccadic eye movements, even in darkness, generate event-related potentials recorded in medial septum and medial temporal cortex. Brain Research 756:168–173.

Spelke E (2000) Core Knowledge. American Psychologist, 55: 1233–1243.

Standing L (1973). Learning 10,000 pictures. *Quarterly Journal of Experimental Psychology*. 25: 207–222.

Stephan H, Bauchot R, Andy OJ (1970) Data on size of the brain and of various brain parts in insectivores and primates. In: Noback C, Montagna W (Eds) Advances in primatology, V.1. Appleton. pp. 289–297.

Stephan H (1972). Evolution of primate brains: a comparative anatomical approach. In: R. Tuttle (Ed), Functional and Evolutionary Biology of Primates, Aldine-Atherton. pp. 155–174.

Stephan H, Baron G, Frahm H (1986) Comparative size of brain and brain components. Comparative Primate Biology, 4: 1–38.

Stephen H, Frahm H, Baron G (1981). New and revised data on volumes of brain structures in insectivores and primates. *Folia Primatologica*, 35: 1–29.

Steriade M (1993) Central core modulation of spontaneous oscillations and sensory transmission in thalamocortical systems. Current Opinion in Neurobiology 3:619–625.

Steriade M (1997) Synchronized activities of coupled oscillators in the cerebral cortex and thalamus at different levels of vigilance. Cerebral Cortex 7:583–604.

Steriade M, Llinas RR (1988) The functional states of the thalamus and the associated neuronal interplay. Physiological Review 68:649–742.

Steriade M, Datta S, Pare D, Oakson G, Curro Dossi RC (1990) Neuronal activities in brain-stem cholinergic nuclei related to tonic activation processes in thalamocortical systems. Journal of Neuroscience, 10:2541–2559.

Striedter G (2005). Principles of Brain Evolution. Sinauer Associates.

Sur M, Angelucci A, Sharma J. (1999) Rewiring cortex: The role of patterned activity in development and plasticity of neocortical circuits. *Journal of Neurobiology* 41: 33–43.

Swadlow HA (1983) Efferent systems of primary visual cortex: a review of structure and function. Brain Research, 287:1–24.

Swadlow HA, Gusev AG, Bezdudnaya T (2002) Activation of a cortical column by a thalamocortical impulse. Journal of Neuroscience, 22:7766–7773.

Swanson L (2002). Brain Architecture: Understanding the basic plan. Oxford University Press.

Szentagothai J (1975) The 'module-concept' in cerebral cortex architecture. Brain Research, 95:475–496.

Tobias P (1985). History of Physical Anthropology in Southern Africa. *Yearbook of Physical Anthropology*, 28: 1–52.

Trevathan, W. (1996). The Evolution of Bipedalism and Assisted Birth. *Medical Anthropology Quarterly*, 10: 287–290.

Trut L (1999). Early Canid Domestication: The Farm-Fox Experiment. *American Scientist*, 87: 160–169.

Tyler L, Stamatakis E, Bright P, Acres K, Abdallah S, Rodd J, Moss H. (2004). Processing objects at different levels of specificity. Journal of Cognitive Neuroscience 16: 351–362.

Ullian E, Sapperstein S, Christopherson K, Barres B (2001). Control of synapse number by glia. Science, 291: 657–661.

Valverde F (2002) Structure of the cerebral cortex. Intrinsic Organization and comparative analysis of the neocortex. Revista de Neurologia 34:758–780.

Vanderwolf CH (1992) Hippocampal activity, olfaction, and sniffing: an olfactory input to the dentate gyrus. Brain Research 593:197–208.

Vargha-Khadem F, Gadian D, Copp A, Mishkin M (2005). FOXP2 and the neuroanatomy of speech and language. *Nature Reviews Neuroscience*, 6: 131–137.

von Bartheld C (1999) Systematic bias in an "unbiased" neuronal counting technique. Anatomical Record, 257:119–120.

von Bartheld C (2001) Comparison of 2-D and 3-D counting: the need for calibration and common sense. Trends in Neuroscience 24:504–506.

von der Malsburg C (1973) Self-organization of orientation sensitive cells in the striate cortex. Kybernetik 14:85–100.

Wade N (2006) Before the dawn. Penguin.

Wallace M, Kitzes L, Jones E (1991) Intrinsic inter- and intralaminar connections and their relationship to the tonotopic map in cat primary auditory cortex. Experimental Brain Research 86:527–544.

Weaver T, Roseman C, Stringer C. (2007). Were neandertal and modern human cranial differences produced by natural selection or genetic drift? *Journal of Human Evolution*, 53: 135–145.

White EL, Peters A (1993) Cortical modules in the posteromedial barrel subfield (Sml) of the mouse. Journal of Comparative Neurology, 334:86–96.

Winer JA, Larue DT (1987) Patterns of reciprocity in auditory thalamocortical and corticothalamic connections: study with horseradish peroxidase and autoradiographic methods in the rat medial geniculate body. Journal of Comparative Neurology, 257:282–315.

Wong H, Liu X, Matos M, Chan S, Perez-Otano I, Boysen M, Cui J, Nakanishi N, Trimmer J, Jones E, Lipton S, Sucher N (2002) Temporal and regional expression of NMDA receptor subunit NR3A in the mammalian brain. Journal of Comparative Neurology, 450:303–317.

Wrangham R (2004) Killer Species. *Daedalus*, 133: 25–35.

Wyss JM, VanGroen T (1995) Projections from the anterodorsal and anteroventral nucleus of the thalamus to the limbic cortex in the rat. Journal of Comparative Neurology, 358:584–604.

Zhang SJ, Huguenard JR, Prince DA (1997) GABAa receptor mediated Cl- currents in rat thalamic reticular and relay neurons. Journal of Neurophysiology 78:2280–2286.

Zhu JJ, Connors BW (1999) Intrinsic firing patterns and whisker-evoked synaptic responses of neurons in the rat barrel cortex. Journal of Neurophysiology 81:1171–1183.

INDEX